Climate Security in the Indo-Pacific

This book explores the intersection of climate change and security in the Indo-Pacific, from the South China Sea disputes to national and regional security practices in India, Japan and Southeast Asia, highlighting the vulnerability of countries facing extreme weather, sea-level rise, and geopolitical tensions.

The Indo-Pacific is the world's most exposed region to natural hazards, and climate change is highly likely to impact the frequency and severity of these events, threatening people's lives and livelihoods. Over the last two decades, research into the security implications of climate change has gained considerable interest among scholars and policy analysts across the world. Evidence-based research has shown how the impact of climate change such as sea-level rise, long droughts and extreme weather events have led to loss of livelihoods, economic downturns, forced migration and health threats from emerging infectious diseases. The effects of climate change in turn have led to greater potential for conflicts breaking out in societies most affected by climate change. As countries in the region face both climate challenges and strategic competition, their experiences offer valuable insights into how we can tackle this issue.

This book is essential reading for students, scholars and policymakers engaged in the fields of climate change, security studies and international relations.

The chapters in this book were originally published as a special issue of *Third World Quarterly.*

Mely Caballero-Anthony is Professor of International Relations, holds the President's Chair for International Relations and Security Studies, heads the Centre for Non-Traditional Security Studies and is Associate Dean for external engagement at the S. Rajaratnam School of International Studies (RSIS) at Nanyang Technological University, Singapore.

Alistair D. B. Cook is Coordinator of the HADR Programme and Senior Fellow at the Centre for Non-Traditional Security Studies, S. Rajaratnam School of International Studies (RSIS) at Nanyang Technological University, Singapore.

ThirdWorlds

Edited by **Shahid Qadir**, *University of London, UK*

ThirdWorlds will focus on the political economy, development and cultures of those parts of the world that have experienced the most political, social, and economic upheaval, and which have faced the greatest challenges of the postcolonial world under globalisation: poverty, displacement and diaspora, environmental degradation, human and civil rights abuses, war, hunger, and disease.

ThirdWorlds serves as a signifier of oppositional emerging economies and cultures ranging from Africa, Asia, Latin America, Middle East, and even those 'Souths' within a larger perceived North, such as the U.S. South and Mediterranean Europe. The study of these otherwise disparate and discontinuous areas, known collectively as the Global South, demonstrates that as globalisation pervades the planet, the south, as a synonym for subalterity, also transcends geographical and ideological frontier.

Marx and Lenin in Africa and Asia
Socialism(s) and Socialist Legacies
Edited by Harry Verhoeven

Affect, Emotions and Power in Development Studies Theory and Practice
Edited by Tanya Jakimow

Migration Politics across the World
Edited by Katharina Natter and Hélène Thiollet

Narrating Working Lives
Oral Histories of Work in Neoliberal Africa
Edited by Jörg Wiegratz, Joseph Mujere and Joost Fontein

Studies of Childhoods in the Global South
Towards an Epistemic Turn in Transnational Childhood Research?
Edited by Afua Twum-Danso Imoh, Lucia Rabello de Castro and Orna Naftali

The Social Life of Skills in the Global South
Edited by Trent Brown and Geert De Neve

Climate Security in the Indo-Pacific
Variations, Contestations and Convergence of Security Practices
Edited by Mely Caballero-Anthony and Alistair D. B. Cook

For more information about this series, please visit:
www.routledge.com/ThirdWorlds/book-series/TWQ

Climate Security in the Indo-Pacific

Variations, Contestations and Convergence of Security Practices

Edited by
Mely Caballero-Anthony and Alistair D. B. Cook

Routledge
Taylor & Francis Group
LONDON AND NEW YORK

Cover image by chuchart duangdaw. Courtesy of gettyimages.

First published 2025
by Routledge
4 Park Square, Milton Park, Abingdon, Oxon OX14 4RN

and by Routledge
605 Third Avenue, New York, NY 10158

Routledge is an imprint of the Taylor & Francis Group, an informa business

British Library Cataloguing in Publication Data
A catalogue record for this book is available from the British Library

ISBN13: 978-1-041-07840-1 (hbk)
ISBN13: 978-1-041-07841-8 (pbk)
ISBN13: 978-1-003-64247-3 (ebk)

DOI: 10.4324/9781003642473

Typeset in Myriad Pro
by Newgen Publishing UK

Publisher's Note
The publisher accepts responsibility for any inconsistencies that may have arisen during the conversion of this book from journal articles to book chapters, namely the inclusion of journal terminology.

Contents

Citation Information

The chapters in this book were originally published in the *Third World Quarterly*, volume 45, issues 14 (2024). When citing this material, please use the original page numbering for each article, as follows:

Introduction

Understanding climate security in the Indo-Pacific
Mely Caballero-Anthony and Alistair D. B. Cook
Third World Quarterly, volume 45, issues 14 (2024), pp. 2039–2046

Chapter 1

Climate security in Southeast Asia: navigating concepts, approaches and practices
Mely Caballero-Anthony
Third World Quarterly, volume 45, issues 14 (2024), pp. 2047–2064

Chapter 2

Climate change in the Asia-Pacific security architecture – the case of ASEAN
Alistair D. B. Cook
Third World Quarterly, volume 45, issues 14 (2024), pp. 2065–2087

Chapter 3

Cascading risks in a social-ecological system: the South China Sea disputes
Maria Ortuoste
Third World Quarterly, volume 45, issues 14 (2024), pp. 2088–2107

Chapter 4

Shifting discourses of climate security in India: domestic and international dimensions
Dhanasree Jayaram
Third World Quarterly, volume 45, issues 14 (2024), pp. 2108–2126

Chapter 5

Climate security and Japan's new national security strategy: a policy analysis
Hideshi Tokuchi
Third World Quarterly, volume 45, issues 14 (2024), pp. 2127–2144

Notes on Contributors

Mely Caballero-Anthony is Professor of International Relations, holds the President's Chair for International Relations and Security Studies, heads the Centre for Non-Traditional Security Studies, and is Associate Dean for external engagement at the S. Rajaratnam School of International Studies (RSIS) at Nanyang Technological University, Singapore.

Alistair D. B. Cook is Coordinator of the HADR Programme and Senior Fellow at the Centre for Non-Traditional Security Studies, S. Rajaratnam School of International Studies (RSIS) at Nanyang Technological University, Singapore.

Dhanasree Jayaram is Senior Assistant Professor, Department of Geopolitics and International Relations, Manipal Institute of Social Sciences, Humanities and Arts, and Co-coordinator, Centre for Climate Studies, Manipal Academy of Higher Education (MAHE), Manipal, India. She is also Research Fellow, Earth System Governance Network; Member, Climate Security Expert Network; Research Fellow, Centre for Public Policy Research, Kochi, India; Non-resident Fellow, Finnish Institute of International Affairs, Helsinki, Finland; Visiting Fellow, Institute of Peace and Conflict Studies, New Delhi, India; and Associate, Network for Education and Research on Peace and Sustainability, Hiroshima University, Japan. She was a Research Fellow at Centre Marc Bloch and Guest Researcher at Freie Universität, Berlin, under the Alexander von Humboldt Foundation's International Climate Protection Fellowship in 2022–2023. She pursued a visiting fellowship (Erasmus Mundus) at Leiden University, the Netherlands, in 2014–2015; and a postdoctoral fellowship at University of Lausanne, Switzerland, under the Swiss Government Excellence Scholarship in 2018–2019. Dhanasree's areas of research include environmental and climate security; climate diplomacy; environmental geopolitics in Southern Asia, Indo-Pacific, and polar regions; gender and climate governance; and geopolitical/security implications of solar geoengineering.

Maria Ortuoste is currently Professor of Political Science at California State University, East Bay. Prior to that appointment, she served for eight years at the Center for International Relations and Strategic Studies which is housed in the Philippines' Foreign Service Institute. Her major research interests are regional maritime security, ASEAN and the ASEAN Regional Forum, the South China Sea, and the domestic politics of Southeast Asian countries. She recently finished several projects: a report titled *The Nexus of Women, Peace and Security and Maritime Security Agendas in ASEAN*; and two book chapters titled "Terrorist threats to security"; and "Philippines in the South China Sea." She is currently working on two projects: "The Manila Connection: Strategic Competition and Economic

Corridors," and "The Social Construction of Maritime Space and Security in Southeast Asia." She is also the program chair for the 2025 conference of the International Studies Association-West.

Hideshi Tokuchi is President of the Research Institute for Peace and Security, a Tokyo-based independent think tank. He received his M.A.L.D. degree from the Fletcher School of Law and Diplomacy, Tufts University, Medford, Massachusetts, United States, in 1986, and retired from the Japanese public service as Vice-Minister of Defense for International Affairs in 2015. His research interests lie in the study of Japan's national security and defence policy, particularly related to international security cooperation and to maritime security. He has been also teaching national and international security issues at a number of universities in Tokyo.

Introduction: understanding climate security in the Indo-Pacific

Mely Caballero-Anthony and Alistair D. B. Cook

ABSTRACT
The Indo-Pacific region has gained traction in security discussions that predominantly focus on traditional security concerns of major power rivalry, statecraft and economic security. Less attention is paid to the impact of climate change on the states and societies in the region most exposed to natural hazards and the knock-on effects that a changing climate poses. Although there are varied and unevenly distributed impacts in the region, the perspectives of those located in the region have not gained sufficient attention in academic debates. This article introduces a volume of *Third World Quarterly* where scholars from within the Indo-Pacific examine the place and priority climate change garners within national and regional security discussions.

Introduction

Over the last two decades, research into the climate change and security nexus has gained considerable interest among scholars and policy analysts across the world. Global climate changes are increasing the intensity and frequency of extreme weather events, long droughts and sea-level rise. Evidence shows how these impacts lead to economic downturns, forced migration and health threats from emerging infectious diseases and livelihood loss. The consequent effects in society result in issues like conflict and societal tensions (International Alert 2007; Gleditsch, Salehyan, and Nordas 2007; Theisen, Gleditsch, and Buhaug 2013). The cultural and economic impacts as well as other effects of climate change aggravate the conditions for conflicts (Caballero-Anthony 2023, 202).

Much research on the conflict–climate change nexus and the climate change–security nexus focus on case studies and a geographic focus on Sub-Saharan Africa, the Sahel region and the Middle East (Calmon, Jacovetti, and Koné 2021; Ide 2017; Nel 2015; Rüttinger et al. 2015; Scheffran, Marmer, and Sow 2012; Vivekananda et al. 2017) with far less attention to the Indo-Pacific region. Where climate security in the Indo-Pacific region is covered it is within the contours of the Global South (Abdenur 2021; Haug, Braveboy-Wagner, and Maihold 2021). We identify this as a significant shortcoming, particularly as countries in the Indo-Pacific constitute the top 10 most vulnerable countries to climate change impacts: Bangladesh, Cambodia, China, India, Indonesia, Laos, Pakistan, Philippines, Myanmar and Thailand. The Indo-Pacific is widely known as the region most exposed to natural hazards, with climate change causing more frequent and intense extreme weather events including typhoons and cyclones, exposure to sea-level rise in archipelagic countries, large coastal population centres and small island states, and extreme heat with significant consequences for densely populated urban areas and large geographic areas (Caballero-Anthony 2023, 202). In this volume the articles cover Northeast, Southeast and South Asia with particular focus on the Association of Southeast Asian Nations (ASEAN), India, Japan and the South China Sea. These countries in the Indo-Pacific are exposed to natural hazards and are also now facing more frequent and extreme weather events very likely caused by climate change. We argue that these events overlay pre-existing and longer-term structural issues, all of which pose significant challenges to the region today such as inequitable economic development, insufficient investment in climate change adaptation and disaster risk reduction, and internal conflicts.

More significant also is the fact that the Indo-Pacific is a geo-strategic arena in terms of the geopolitical tensions and competition between major powers. The region comprises both large and small nation-states and societies from India and Japan to the littoral states of the South China Sea and wider ASEAN member states, an arena for competition and cooperation between China and the United States, and host to a patchwork of minilateral and multilateral frameworks like ASEAN. Against this background, this volume argues that conceptualisations of climate security in the Indo-Pacific merit an in-depth examination of the extent to which the notion of climate security has found traction in the security discourses among states and the extent to which this is defining the nature of intra- and inter-state relations in the region. We contend that the analysis of climate security as reflected in the current literature, particularly in peace and conflict studies, is not path-dependent, particularly for a region where ideas of comprehensive security and non-traditional security go beyond the more state-centric and realist understanding of security. Instead, how climate security is framed, the attention on the kinds of threats to expect from climate change and their impact on security referents – state, individuals, economy – and the nature of solutions to the problem, among other factors, allow for better understanding of how climate security is 'received' and defined by state and non-state actors in the Indo-Pacific.

Aims and scope

Since the United Nations Security Council open debate in 2007 on the impact of climate change on peace and security, many countries in 'the West' have included climate change as part of their national security strategies. Beyond the corridors of power in 'the West', there remains limited coverage of the ways in which climate change and security are understood and acted upon in the Indo-Pacific region. Climate change is often seen in policy documents and security assessments as an avenue for cooperation or a threat to stability. As mentioned earlier, despite the fragility risks faced by states in the Indo-Pacific, the place of climate change in the security discourse of the region remains peripheral.

As a result, we first hosted a virtual roundtable on 'Climate Security in the Indo-Pacific: Implications for Defence and Foreign Affairs' on 25 October 2021 to explore the impacts and parameters of climate change in the region. After a fruitful discussion, we followed up on 2 November 2022 with an in-person workshop hosted by the S. Rajaratnam School of International Studies, Nanyang Technological University in Singapore. This volume is the product of the papers delivered at the workshops.

This volume therefore seeks to go some way to contribute to a wider conversation on climate change and security in the pages of *Third World Quarterly*. We include voices from across the Indo-Pacific, including scholars based in South, Southeast and Northeast Asia, a combination of male and female scholars at different career levels, and a variety of methodological approaches. This offers a pioneering study that opens space for different views on climate change and security in the Indo-Pacific. Bringing these contributors together in a single volume offers insight into the place, understanding, challenges and prospects for climate security. The volume considers whether, how and why climate change is understood in security discourse and the impacts of climate change on the security of countries and the region.

This collection answers, to what extent has the concept of climate security been accepted in the security discourses in the Indo-Pacific? How is climate change and security framed in the region? Has climate change influenced the way security is understood and approached in the region? What are the policy implications of these approaches? Do states and/or regional organisations subscribe to the 'multiplier effect' that climate change is a trigger for, and significantly contributes to, conflict? Or would a regional climate security agenda offer opportunities for more dialogue and cooperation?

The inherent diversity of the Indo-Pacific region lends to different approaches in understanding the nexus between climate change and security. This region's geopolitical heterogeneity – environmental vulnerabilities, varying government structures and security frameworks – challenges a singular view or security approach in dealing with climate insecurities. While traditional security paradigms which focus on state-centric and military threats dominate in debates at the global level, these are proving to be insufficient in capturing the complex, non-linear risks posed by climate change, such as resource scarcity, food insecurity, forced migration and environmental degradation, among others. And while a human-centred approach to dealing with climate change is arguably compelling, significant tensions exist between different actors on how best to respond to its impacts, while deep-seated concerns about state sovereignty and non-intervention persist.

The volume presents a platform for the voices of those based in a region most exposed to climate change impacts and to understand how these are similar or different to those in other parts of the world. Its central contribution to the conversation on climate change and security is to open it up to contestation from those most adversely affected by climate change while advancing a better understanding of its complex effects on states and societies across different regions.

Climate change and security in the Indo-Pacific

In 'Climate Change in Southeast Asia: Navigating Concepts, Approaches and Practices', Mely Caballero-Anthony argues that while considered to be rarely a primary cause of conflict, climate change is a risk multiplier and an aggravating factor for instability and conflict as it can exacerbate underlying vulnerabilities and compound existing grievances. Climate change can also undermine efforts and initiatives for conflict prevention and sustaining peace. However, despite significant increase in multilateral discussion on the impact of climate change on global peace and security and the growing number of organisations like the European Union, the African Union and the Pacific Island Forum advancing the agenda of climate security, it has emerged as a contentious issue. Asian countries like China and

India and in Southeast Asia have been lukewarm, at best, to the climate security framing. The article examines the reasons why climate security is not finding enough traction in Southeast Asia and why concepts like comprehensive security and non-traditional security are the preferred lexicon by states in the region. Her article argues that the reticence by ASEAN member states to embrace the term 'climate security' stems less from a denial of climate change's security implications and more from concerns that framing climate issues in security terms could heighten focus on regional conflicts, challenge notions of sovereignty and non-intervention and contribute to the stigmatisation of less-developed countries. She also contends that with the geopolitical tensions in the wider Indo-Pacific region, there are worries that linking climate change with security increases the risks of more military involvement in non-military issues, potentially causing more competition than cooperation. However, she further argues that ASEAN's reluctance to adopt the term 'climate security' risks not only downplaying the criticality of the issue but also weakening the urgency to build regional capacity for anticipatory climate action including the ability to effectively respond to potential climate-induced conflicts at the national and regional level (Caballero-Anthony 2024, in this volume).

Considering the place of climate change in the Asia-Pacific security architecture, Alistair D. B. Cook explores the developments and discourse in ASEAN. The article reviews the official statements, declarations, and communiques related to climate change from 2015 to 2023. It reveals a growing acknowledgment of climate change and its impact on peace and security within the regional context. Leaders are signalling the emergence of a mindset shift through recent official documents and speeches that climate change has a place within regional security discussions. These discussions remain inconclusive on how best to overcome an atomised approach to the security impacts of climate change. Most substantive discussions take place largely in hybrid forums, despite official recognition in traditional platforms, which may undermine efforts for a more inclusive regional discourse on climate security in ASEAN. These recent developments highlight greater but cautious engagement by regional elites with climate change as an issue impacting peace and security within the Asia-Pacific security architecture (Cook 2024, in this volume).

Turning to the South China Sea disputes, Maria Ortuoste explains how global political and economic structures shaped policies and approaches by states that have had cascading risks furthering environmental degradation and conflict at different levels of society for much of the past century. She explores this through territorialisation, militarisation, extraction and exploitation, framing the South China Sea disputes as a social-ecological system, and identifies these ultimately as a wicked problem where issues cross state boundaries and domains, and where stakeholders range from fishers to governments to shipping industries. Instead of drawing on securitisation, the article draws on 'riskification', which proponents view as having a greater focus on governance than on military action. Thus, it is particularly important to emphasise stewardship rather than ownership when it comes to discussions on any code of conduct in the South China Sea. This will allow the discussions to be climate-centred, rather than threat-centred. Ortuoste argues that shifting the narrative from security to risk is a better way to address the very real risk of the collapsing biological subsystem upon which the social subsystem ultimately depends (Ortuoste 2024, in this volume). This article highlights the reluctance to utilise 'climate security' as a framework for analysis because 'national security' dominates rather than 'human security' in the context of the South China

Sea disputes. It does beg the question, how, if at all, does the grammar of riskification differ from that of comprehensive security?

Moving beyond Southeast Asia to South Asia, Dhanasree Jayaram explores the changing debates of climate security in India. In her article she finds these debates are largely conditioned by developmental priorities within the domestic context and geopolitical pressures in international interactions, which are not necessarily mutually exclusive. The confluence of factors influencing national-level decision makers essentially comes from within their localities as well as international pressure. In this context, her article investigates the discussion of climate security in India through considering a range of different actors from academia and think tanks, civilian governments, militaries and non-governmental organisations (Jayaram 2024, in volume). Dhanasree argues that climate change has been gradually integrated into the domestic security policymaking agenda in tandem with the growing recognition of worsening climate vulnerabilities in India. However, while this has occurred domestically, India remains reluctant to engage with the existing climate security discourses at the international level (Jayaram 2024, in this volume). These observations chime well with those in Southeast Asia that there are concerns not on climate science, impacts or exposure but rather more fundamental concerns over power, sovereignty and agency with the invocation of climate security as articulated in international forums.

Regarding Northeast Asia, Hideshi Tokuchi analyses climate security in Japan's new national security strategy. Through his investigation, the policy developments by an Asian state illustrate the motivations of governments in the Indo-Pacific to respond to the challenges posed by climate change while concurrently noting the need to engage international partners. He finds that the Japanese government is more sensitive to climate security concerns but its policies remain framed by geopolitical and traditional security issues (Tokuchi 2024, in this volume). He argues that policy developments on climate security in Japan need to be integrated as part of the government's overall international cooperation with a particular focus on its military partnerships. These efforts need to pay particular attention to improving climate change adaptation and response efforts of militaries (Tokuchi 2024, in this volume). This falls in line with recent observations in the literature that highlight the rapid and widespread institutionalisation of climate security with an emerging geopolitical separation between countries in the Global North and Global South (Hardt et al. 2024). Hideshi Tokuchi sees Japan's efforts in this regard contributing to the international efforts to build a new security architecture in the Indo-Pacific.

Of particular note are the activities of Japan's Ministry of Defense, which released its Response Strategy on Climate Change. He finds that this strategy demonstrates the defence ministry's commitment to integrate and recognise climate change impacts on national security. As it is quite new for Japan, he argues that there is much room for expansion in response, adaptation and reduction efforts. Further, he contends that concerns over climate security are not an attempt to pursue covert political goals but that scientific integrity should be respected and acted upon. In his conclusion he suggests that Japan's international cooperation on climate security policy development must be promoted to advance climate security policy at home in Japan (Tokuchi 2024, in this volume). Through the different analytical lenses utilised in this volume there are several key concerns but also significant developments as states in the region grapple with the climate change impacts on peace and security in the Indo-Pacific region.

Implications for policy and research

Through these articles, this volume on climate security in the Indo-Pacific contributes important insights into the place and priority of climate change in security discussions in the region. Despite the debates over how states and societies should deal with the security challenges brought on by climate change, it is recognised as a collective action problem with significant opportunities for cooperation. Climate change, as a transnational challenge, necessitates collaborative solutions in areas across the disaster management cycle from prevention, mitigation, preparedness, response and recovery to rehabilitation and 'building back better' for a more sustainable planet for states and societies. Climate security also underscores the imperatives of strengthening capacities to address cross-cutting threats posed by food and water insecurity, health crises, and environmental degradation.

As identified in this volume, there are atomised approaches to climate change, peace and security worthy of further investigation in terms of the growing differences between the Global North and Global South, the localised impacts and leadership in different countries, the emerging discussions at the regional level, the intersection of cross-cutting issues, and the increasing priority of climate change by militaries. More importantly, the design and inclusion of people in policymaking processes, the role of expertise, and the emergence of new avenues of cooperation and conflict need further exploration and discussion of how and why the debates over climate change are increasingly shaping global peace and security. At the same time, highlighting a coordinated global response is essential in addressing the array of complex threats posed by climate change.

The latest Multilateralism Index 2024 highlights that while engagement is up globally across all domains except trade, there is deterioration in performance in peace and security, and there are mixed trends in climate action (IPI-IPE 2024). As hosts to several multilateral frameworks, the Indo-Pacific provides several case studies of cooperative efforts underway that aim to mitigate climate risks and illustrate how a shared threat can also serve as a platform to manage conflicts, build trust and foster inclusive leadership. The pages of *Third World Quarterly* are a platform to exchange views and share research. It is important for scholars and practitioners alike to navigate the climate change impacts on peace and security and provide insights from people and places that are often overlooked and underappreciated in this endeavour. To be sure, understanding how climate change influences the security dynamics of states in different regions around the world is crucial, as this will shape the contours of international relations in a climate-changed world. We hope the articles in this volume contribute to these discussions and provide several views from the Indo-Pacific that spark wider reflection and consideration in the broader global discussions on climate security.

Acknowledgements

First and foremost, we thank the participants of the virtual and in-person workshops for making them a fruitful exchange of views and sharing perspectives. The workshops would not have been possible without the administrative and funding support by the S. Rajaratnam School of International Studies, Nanyang Technological University, Singapore.

Disclosure statement

No potential conflict of interest was reported by the author(s).

ORCID

Mely Caballero-Anthony http://orcid.org/0000-0003-1064-745X
Alistair D. B. Cook http://orcid.org/0000-0003-1934-5474

Bibliography

Abdenur, A. E. 2021. "Climate and Security: UN Agenda-Setting and the 'Global South'." *Third World Quarterly* 42 (9): 2074–2085. https://doi.org/10.1080/01436597.2021.1951609.

Caballero-Anthony, M. 2023. "Peace and Security Studies in Southeast Asia in a Changing Security Environment." *Asian Journal of Peacebuilding* 11 (1): 193–212. https://doi.org/10.18588/202305.00a346.

Caballero-Anthony, M. 2024. "Climate Security in Southeast Asia: Navigating Concepts, Approaches and Practices." *Third World Quarterly* 1–18. https://doi.org/10.1080/01436597.2024.2370353.

Calmon, D., C. Jacovetti, and M. Koné. 2021. "Agrarian Climate Justice as a Progressive Alternative to Climate Security: Mali at the Intersection of Natural Resource Conflicts." *Third World Quarterly* 42 (12): 2785–2803. https://doi.org/10.1080/01436597.2021.1965870.

Cook, A. D. B. 2024. "Climate Change in the Asia-Pacific Security Architecture – The Case of ASEAN." *Third World Quarterly* 1–23. https://doi.org/10.1080/01436597.2024.2388291.

Gleditsch, N. P., I. Salehyan, and R. Nordas. 2007. *Climate Change and Conflict: The Migration Link*. New York: International Peace Institute. 1–13.

Hardt, J. N., D. Jayaram, C. Harrington, D. McLaren, N. P. Simpson, A. D. B. Cook, M. C. Oliveira, et al. 2024. "The Challenges of the Increasing Institutionalization of Climate Security." *PLoS Climate* 3 (4): e0000402. https://doi.org/10.1371/journal.pclm.0000402.

Haug, S., J. Braveboy-Wagner, and G. Maihold. 2021. "The 'Global South' in the Study of World Politics: Examining a Meta Category." *Third World Quarterly* 42 (9): 1923–1944. https://doi.org/10.1080/01436597.2021.1948831.

Ide, T. 2017. "Space, Discourse and Environmental Peacebuilding." *Third World Quarterly* 38 (3): 544–562. https://doi.org/10.1080/01436597.2016.1199261.

International Alert. 2007. A Climate of Conflict: The Links Between Climate Change, Peace and War. https://www.international-alert.org/publications/climate-conflict/

IPI-IPE (International Peace Institute and Institute for Economics and Peace). 2024. "Multilateralism Index 2024." New York, October. https://www.ipinst.org/wp-content/uploads/2024/10/2410_Multilateralism-Index-2024.pdf.

Jayaram, D. 2024. "Shifting Discourses of Climate Security in India: Domestic and International Dimensions." *Third World Quarterly* 1–19. https://doi.org/10.1080/01436597.2024.2314003.

Nel, A. 2015. "The Neoliberalisation of Forestry Governance, Market Environmentalism and Re-Territorialisation in Uganda." *Third World Quarterly* 36 (12): 2294–2315. https://doi.org/10.1080/01436597.2015.1086262.

Ortuoste, M. 2024. "Cascading Risks in a Social-Ecological Wisystem: The South China Sea Disputes." *Third World Quarterly* 1–20. https://doi.org/10.1080/01436597.2024.2357658.

Rüttinger, L., G. Stang, D. Smith, D. Tänzler, J. Vivekananda, A. Carius, O. Brown, et al. 2015. "A New Climate for Peace – Taking Action on Climate and Fragility Risks." Executive Summary. Berlin: adelphi; London: International Alert; Washington: The Wilson Center; Paris: EUISS. https://climate-diplomacy.org/sites/default/files/2020-11/NewClimateForPeace_FullReport_small_0.pdf

Scheffran, J., E. Marmer, and P. Sow. 2012. "Migration as a Contribution to Resilience and Innovation in Climate Adaptation: Social Networks and Co-Development in Northwest Africa." *Applied Geography* 33 (1): 119–127. https://doi.org/10.1016/j.apgeog.2011.10.002.

Theisen, O. M., N. P. Gleditsch, and H. Buhaug. 2013. "Is Climate Change a Driver of Armed Conflict?" *Climatic Change* 117 (3): 613–625. https://doi.org/10.1007/s10584-012-0649-4.

Tokuchi, H. 2024. "Climate Security and Japan's New National Security Strategy: A Policy Analysis." *Third World Quarterly* 1–18. https://doi.org/10.1080/01436597.2024.2314681.

Vivekananda, J., S. Fetzek, M. Mobjörk, A. Sawas, and S. Wolfmaier. 2017. "Action on Climate and Security Risks." *Review of Progress 2017*. Clingendael: Den Haag, Netherlands.

Climate security in Southeast Asia: navigating concepts, approaches and practices

Mely Caballero-Anthony

ABSTRACT
Climate security is gaining significant traction in global discourses on how to respond urgently to the challenges brought on by climate change. In Southeast Asia, however, the term 'climate security' raises concerns about the unintended consequences of securitisation to frame the complex and cross-cutting impact of climate change. This paper examines these concerns against the evolving discourses on climate security. It argues that the Association of Southeast Asian Nations (ASEAN's) reservations about invoking the language of security on climate change are underpinned by concerns about sovereignty and non-interference in internal affairs. The paper also argues that ASEAN's hesitance to adopt 'climate security' downplays the critical need to build regional capacity for anticipatory climate action, including the ability to respond to potential climate-induced conflicts.

Introduction

According to the Sixth Assessment Report by the Intergovernmental Panel on Climate Change, climate change is rapid and intensifying in every region in the world (IPCC 2023). This report, like the others before it, describes the multi-faceted and cross-cutting impacts of climate change on the well-being and security of states and societies around the globe.

Referring to the report, United Nations Secretary-General Antonio Gutierrez declared a 'climate emergency' facing humanity (UN 2022). International non-governmental organisations (NGOs) have implored leaders to, as the World Wildlife Fund (2023) put it, 'heed the science and act immediately with the pace and scale necessary to decarbonise our economies in time … [noting that] we are running out of time, and no countries are yet on track for a 1.5 °C pathway'. More recently, 18 national governments and the European Union[1] have declared a climate emergency (Climate Emergency Declaration 2023).

That only 18 out of 193 countries have made the declaration despite the scientific evidence reflects a lack of urgency across much of the international community. One plausible explanation for this is that unless climate change is securitised, the exigency to act is absent (Buzan, Wæver, and de Wilde 1998). As several studies show, securitisation can generate greater attention and urgency in addressing security threats in Asia (Caballero-Anthony,

Emmers, and Acharya 2006; Emmers, Caballero-Anthony, and Acharya 2006; Haacke and Morada 2010). Thus, it could be argued that unless the term 'climate security' is adopted in the discourses of states, and specifically in the security lexicon used by the policy community and security establishments, there will be no urgency to respond.

The international discourse on climate security is a relatively recent development. It was not until 2007 that climate change was raised at the United Nations Security Council (UNSC) by the United Kingdom as the greatest threat to international peace and security. Although eight UNSC thematic debates on climate change have been held since 2007, a UNSC resolution has yet to materialise in the face of major opposition from China, India and Russia (Scartozzi 2022), who were not so much questioning whether climate change is a threat to humanity but more 'what role the Security Council should play in addressing this threat, and under what circumstances' (Security Council Report 2022, 3).

Other reasons for the lack of consensus on climate security include the fundamental question of whether climate change is a development issue or a security issue, and the contention that the climate security agenda is mostly driven by Western countries, which raises questions about the motivations behind the securitisation of climate change (McDonald 2023). There has also been disquiet over the unintended consequences of securitisation, including greater military involvement in non-military threats (Eckersley 2007; Hulme 2011; Scott 2008, quoted in McDonald 2013).

These concerns have found resonance in Southeast Asia,[2] where there is a noticeable lack of official discourse on climate security in the Association of Southeast Asian Nations (ASEAN).[3] This is despite the region being considered one of the most vulnerable to climate change. Over the past 50 years, close to 450,000 people have lost their lives and around 525 million people have been impacted by natural hazards in Southeast Asia (ESCAP 2022). The 2021 ASEAN State of Climate Change Report notes that the region is vulnerable to the 'growing intensity and magnitude of extreme weather events, and increasing economic, environmental, and social damage' (ASEAN 2021, xiv).

Countries in Southeast Asia not only face huge challenges in dealing with the geophysical effects of climate change but are also hampered by lower adaptation capacity, lower economic development, and governance issues. That development impacts such as food insecurity, health, and natural disasters have been a focus of discussions in ASEAN is thus not surprising. However, the fact that there have been few to no mentions of possible security threats and instability arising from climate-induced conflicts bears further examination. This reluctance to frame climate change as a security issue may be better understood within the context of the region's preference for the framework of comprehensive security, which guides its security strategies and practices.

Against this background, the objectives of this paper are two-fold. The first is to examine the discourses on climate change in Southeast Asia and analyse the reticence of ASEAN in invoking climate security in its regional discourses. The second is to assess how the differences in presenting the issue of climate change inform the kinds of approaches and policies adopted in the region.

This paper does not aim to provide exhaustive content analyses of the climate security statements by ASEAN states. But given that the global trend of framing climate change as an issue of international peace and security is fairly recent, a snapshot of statements from political leaders and official representatives in ASEAN may provide useful insights on the regional thinking about climate change and security. Regional discussions on climate security could have significant bearing on global efforts to manage the complex security impacts of climate change.

The paper sets forth two arguments. First, ASEAN reservations about adopting the term 'climate security' are underpinned by concerns that such language will draw more attention to regional conflicts and impact sovereignty and non-interference in internal affairs. There is also concern that focusing on conflict-related security challenges will expose the lack of regional capacity to deal with humanitarian crises. Moreover, given the geostrategic dynamics and geopolitical tensions in the Indo-Pacific region, characterised by major power rivalry, there are worries that linking climate change with security increases the risks of growing military involvement in non-military matters, potentially causing more inter-state competition rather than cooperation.

Second, the reluctance of ASEAN to adopt the term 'climate security' risks downplaying the critical need to build regional capacity for anticipatory climate action (Mosello, Rüttinger, and Sauerhammer 2020), including the ability to respond to potential climate-induced conflicts at the national and regional level. Failure to address climate security imperils ASEAN's overarching goal of creating a safe, just and secure regional community.

The paper is organised in six parts. Following the introduction, there is a review of how the climate security concept has evolved, drawing on studies that examine the causal linkage between climate and security. The next section proceeds to examine how the different framings of climate security shape the policy responses advocated by different policy and academic communities. This is followed by a section situating ASEAN's discourses within the ongoing discussion on climate security, and analysing the pathways preferred by countries in the region for dealing with climate-related security risks. Next, this paper identifies factors that explain ASEAN's reservations about invoking climate security. The paper concludes with thoughts on how ASEAN can comprehensively address the complex challenges of climate change without privileging the development dimension over other issues of peace and security.

Unpacking the conceptualisation of climate security

In examining why climate change is a security issue, it is useful to refer to the Copenhagen School's theory of securitisation. Buzan, Wæver, and de Wilde (1998, 21) posit that what makes an issue an international security concern is if it is presented as posing an existential threat to a designated referent object. The process of securitisation is referred to as a 'speech act' in language theory – the utterance of the issue is the act (Buzan, Wæver, and de Wilde 1998, 26). Thus, labelling an issue as involving 'security' allows for it to be surfaced and prioritised.

McDonald (2013, 42–51) traces the emergence of climate security to the 1992 United Nations Framework Convention on Climate Change (UNFCCC) although there had been earlier scholarly works highlighting the scale and scope of climate change and arguing that this threat to the long-term survival of humanity deserves priority and funding. As Oels (2012) notes, the security implications of climate change originated in the late 1980s with the debates raised by NGOs as well as the academic community about the possibility of 'environmental refugees'.

Just a few years later, evidence began to emerge on the specific pathways in which climate change could lead to environmentally induced conflicts, with serious implications for international security. Gleick (1993) observes that climate change affects water resources and could lead to 'water wars' over access to supply of fresh water. Myers (1995) highlights that climate change could interact with socio-economic issues to destroy livelihoods, triggering large numbers of environmental refugees.[4]

Through the 1990s, while debate continued over the causal mechanisms,[5] consensus grew within academic circles that climate change would become a security issue. By the 2000s, discussions of climate security gradually expanded to include policymakers and security establishments as the wide-ranging effects of climate change became more obvious. Three broad frames of climate security began to emerge, each with its own emphasis.

The first saw climate security through the lens of national security. This view came primarily from national security establishments and related think tanks, and much of the emphasis was on emerging threats that militaries need to respond to (Matthew 2000; McDonald 2013). Early expressions of such a view include the landmark 2007 report from the CNA Corporation on the effect of climate change on the national security of the United States, and the 2009 Defence White Paper from the Australian government. These reports essentially suggest that climate change is likely to have a two-pronged effect. First, weather effects are likely to become more frequent, creating new demands on militaries, such as disaster relief and humanitarian aid operations abroad (Australian Government, Department of Defence 2009). Second, climate change could become a 'threat multiplier for instability' at home and abroad (CNA Corporation 2007, 5–7). The notion of threat multiplier has generated more attention to the cross-cutting effects of climate change that could lead to competition, conflict and threats to national security.

Contemporary research on climate security has further explored linkages between climate change and conflict (von Uexkull and Buhaug 2021). This includes how climate change could indirectly give rise to conflict due to its effect on the economy, agriculture and the well-being of communities (Maystadt and Ecker 2014; McGuirk and Burke 2020). Other scholars have deliberated on the causal links between climate-induced migration and conflict (Boas et al. 2019; Brzoska and Fröhlich 2016). Several studies have looked at the interaction of climate impacts with other factors that exacerbate vulnerability, such as resource- or agriculture-dependent livelihoods (Bretthauer 2015; Salehyan and Hendrix 2014) and population loss and growth (Breckner and Sunde 2019).

Given the multiplicity of climate-related issues, von Uexkull and Buhaug (2021) have proposed that future research focus on developing more scenario-based projections to predict climate effects on conflict risk; examining effective policy responses in relation to peacebuilding and climate change; and measuring the security risks and benefits of climate change adaptation and mitigation efforts (see also Busby 2021, quoted in von Uexkull and Buhaug 2021; Witmer et al. 2017).

A second conception of climate security is through the lens of human security. This discourse emerged with the publication of the 1994 Human Development Report by the United Nations Development Programme (UNDP). The report highlights the salience of environmental security, focusing on how damage to the environment jeopardises the core tenets of human security – that is, the individual's 'safety from chronic threats' and 'protection from sudden and hurtful disruptions in the patterns of daily life' (UNDP 1994, 22–24). This focus on the 'security of individuals and communities' would later be endorsed in a UN (2009) report by its Secretary-General on the security implications of climate change.

Later work within this paradigm expanded the definition of environmental security to include the concept of the 'capacity to manage stresses to people's needs, rights, and values' that have been precipitated by environmental change (Barnett, Matthew, and O'Brien 2010, 18). Some of the human security issues related to climate change include how migration dynamics are affected by weather-related events, discussions of climate refugees, and large global estimates of populations being at risk of being displaced (Gemenne 2011).

The third conception of climate security is through the lens of international security – that is, how and why climate change poses a threat to the international community. According to this perspective, responding to the complex climate threats requires internationalism and global cooperation. International organisations are important as coordinators of climate mitigation and adaptation efforts (McDonald 2013) while the tools for the cooperative resolution of issues are seen to come from international institutions, laws and agreements. Examples include the UN Declaration on the Environment and the emerging principles of international environmental law (UNEP n.d.-b).

The 2006 Stern report epitomises this framing in its call for immediate global action on climate change. The report stresses the importance of multilateral frameworks such as the Kyoto Protocol and international organisations such the International Energy Agency (Stern 2006, 458–464). Other studies have looked at how the European Union, the Food and Agriculture Organization of the United Nations (FAO) and the World Bank can help strengthen collective action on climate change (Dellmuth et al. 2018). More recently, a 2022 report from the Stockholm International Peace Research Institute (SIPRI) has urged more ambitious and effective international cooperation and peacebuilding to promote climate security (Black et al. 2022, 20–24; see also Rasmussen and Birk 2012).

Significantly, several scholars point to the UN system as a possible coordinator for climate policy. Purvis and Busby (2004) argue that the UN Secretary-General has an important role in raising the 'visibility of climate change' and 'overcoming obstacles to emissions mitigation' and they suggest creating 'formal institutional mechanisms' to generate consistent attention. Rasmussen and Birk (2012) argue that the UNSC is the only institution in a position to react to global environmental threats to security (while also acknowledging that embargoes and military interventions are unlikely to be solutions to the climate crisis).

The drive for greater collective action on climate security has seen this issue elevated to the UNSC, where, for example, several African states have articulated their experience with the wide-reaching impacts of climate change on stability (of states) and security (Security Council Report 2022). Such UNSC discussions represent possible entry points and pathways for UN action to anticipate, mitigate and manage climate-related security risks in a comprehensive manner. In fact, since 2018, the UNSC has integrated climate change language in resolutions with peace operation mandates. The UN has also established a Climate Security Mechanism that 'supports field missions, UN Resident Coordinators and regional organizations to conduct climate security risk assessments and risk management strategies' (UNEP n.d.-a).

Policy approaches and responses to climate security

The varying discourses on climate security reflect distinct policy directions put forward by the different stakeholders to address the impacts of climate change on security. The most visible are the policies advocating for mitigation and adaptation.

Mitigation focuses on reducing carbon emissions, with much attention being given to advancing the low-carbon energy transition. To advance this agenda, the principle of common but differentiated responsibility was adopted at the 1992 UNFCCC Earth Summit to encourage developed countries to take the lead in decarbonising their energy systems while also helping developing countries to do so both financially and technically. This principle was reiterated at the 15th UN Climate Change Conference of the Parties (COP15) (UNFCCC n.d.).

Adaptation, considered a human-centred approach, focuses on the solutions and actions needed to address the kinds of vulnerabilities faced by communities affected by the impacts of climate change (such as water and food insecurity, disease outbreaks, and forced migration), particularly those from less-developed areas that are at greatest risk due to 'both higher sensitivity and lower adaptive capacity' (IPCC 2007, 781). Several studies suggest that adaptation helps with building climate resilience, with resilience now gaining ground in development, security and disaster management (Boas and Rothe 2016; Methmann and Oels 2015).

While the two mainstream approaches of mitigation and adaptation continue to dominate discourses on climate change, the ones that are garnering more interest, especially among security and defence establishments, are those related to humanitarian assistance and disaster relief (HADR), conflict prevention, and peacebuilding. Militaries are increasingly stepping up to provide transnational HADR during catastrophic natural disasters (Newby 2019; Niazi 2022) due to the capacity limitations of civilian agencies when it comes to providing adequate, timely response in the face of sudden-onset and large-scale emergencies and disasters (Canyon, Ryan, and Burkle 2017; Hofmann and Hudson 2009).

Meanwhile, the push to include climate security in conflict prevention and peacebuilding efforts is built on the premise that countries with the highest levels of ecological threat are statistically likely to be those where peace is at its most tenuous (Black et al. 2022). Krampe, Tarif, and Mobjörk (2020, 2) argue that addressing this security issue implicates a range of policy areas, from foreign policy to defence, development, economic, humanitarian and environmental policy. To better address climate change risks, they advocate investigating how climate change translates into human security risks and under what conditions it does so. Similar studies argue that climate change creates risk multipliers for communities (Day and Caus 2020), and that more effort should go into examining the drivers of conflict to build foresight capacity, especially for the UN, which in turn could be integrated into and significantly improve UN peacebuilding programmes (Black et al. 2022; Hardt and Scheffran 2019).

Situating the climate security discourse in Southeast Asia

Against the evolving security framings of climate change, discourses that invoke the term 'climate security' are hardly found in Southeast Asia. This is despite states, international organisations and non-state actors such as NGOs, civil society organisations (CSOs) and epistemic communities having securitised climate change in various ways. Given that discourses can shape and reflect how climate change impacts are perceived, it is useful to explore how the region's overarching notion of comprehensive security influences perspectives and policy considerations regarding climate security.

Since its articulation in the 1980s, comprehensive security has remained the core concept of security in Southeast Asia. Unlike the conventional notion of security, which focuses mainly on defending state borders from military attack, comprehensive security offers a much broader conceptualisation: it goes 'beyond (but does not exclude) the military threats to embrace the political, economic, and socio-cultural dimensions' (Alagappa 1988, 624). The thinking is that security is multi-dimensional, and states cannot be secure until their economic, political and socio-cultural challenges are addressed. The principles of comprehensive security recognise economic security as fundamental to state security, thereby ensuring political stability and social cohesion. Thus, adopting comprehensive and cooperative strategies to address the array of security challenges confronting the region is integral to Southeast Asian security practices. This entails employing all available approaches – socio-economic, political, security – to guarantee the protection, well-being and security of peoples and states.

It is often challenging to find where climate change sits in ASEAN within such an expansive framing. The complexity and multi-dimensionality of the effects of climate change have made it more difficult to find a definitive space where climate insecurities are discussed in a focused and targeted way. Policy discussions in ASEAN on climate change are fragmented, as seen from the discourse under the three pillars of the ASEAN Community, namely the ASEAN Political-Security Community (APSC), the ASEAN Economic Community (AEC) and the ASEAN Socio-Cultural Community (ASCC).

Within the ASEAN Community framework, mentions of climate change are predominantly found under the socio-cultural pillar, where it is interspersed among discussions on the environment, haze, migration, health, and disaster management and humanitarian assistance (see ASCC 2009). In the 2025 ASCC Blueprint, ASEAN committed to working towards 'a resilient community with enhanced capacity and capability to adapt and respond to social and economic vulnerabilities, disaster, climate change, and other new challenges' (ASCC 2016, 3).

While climate change is featured in many of areas of cooperation under the socio-cultural pillar, most climate-related work is done under the ambit of the ASEAN Ministerial Meetings on the Environment and the ASEAN Senior Officials' Meeting on the Environment. And, despite acknowledging the cross-cutting impacts of climate change, the work of the ASEAN Working Group on Climate Change, whose mission is to promote regional cooperation on climate change, is still limited to the environmental mandate.

Within the ASEAN Community's economic pillar, discussion on climate change occurs largely within the context of how to enhance energy security through advancing renewables and the green economy (AEC 2015). The latter essentially means promoting low-carbon friendly infrastructure and business to help stimulate economic development (EU–ASEAN Business Council 2023). Discussions on the impact of climate change on food security against concerns of food production and supply are found in the agriculture and forestry sectors.

All in all, climate issues are discussed in at least 13 ASEAN sectoral bodies, but it is clear that more needs to be done. The 2020 State of Southeast Asia Survey found that while 52.7% of respondents see climate change as posing an immediate threat to their country's well-being, 52.4% feel that governments have not paid it adequate attention (Tang et al. 2020). Analysts have argued that ASEAN's fragmented approach has not been adequate to the task, and that there is a need to establish an overarching coordinating body dedicated to climate change to ensure a more effective response across the ASEAN Community's three pillars (Seah 2022; Seah and Martinus 2021).

The observable tendency to integrate climate change within the ASEAN Community's socio-cultural pillar, and for discussions to be mainly among environmental, health and disaster management officials, suggests a regional reluctance to bring this issue to, or elevate it within, the political and security arena. Noticeably, mentions of climate change within the political-security pillar are found in discussions to promote regional cooperation on non-traditional security challenges such as managing maritime-related issues and protecting the maritime ecosystem (APSC n.d.). Although climate issues are discussed at political-security and defence platforms such as the ASEAN Defence Ministers' Meeting (ADMM), ADMM-Plus and the ASEAN Regional Forum, the policy thrust has been on increasing cooperation on HADR. Visibly missing are discussions on the potential of climate change to cause instabilities from competition for resources, forced displacements and severe humanitarian emergencies that may require military action. Nor have there been discussions on how climate effects can compound existing internal conflicts, necessitating regional action on conflict prevention and peacemaking efforts.

While preparedness is critical in mitigating the devastating impact of natural disasters, there is hardly any discussion on how the military establishment can proactively deal with climate-induced threats to military bases and assets and future military deployments. Cook and Nanthini (2023) have argued that militaries in the region need to go beyond HADR and turn their attention to national security strategies that consider the emerging riskscape. A sub-text to this riskscape is how climate change could exacerbate geostrategic rivalries that could spill over into conflict and result in a more volatile security environment (Caballero-Anthony et al. 2023).

'Climate-related security' instead of 'climate security'?

Although the term 'climate security' may have been absent in regional discussions and ASEAN documents, 'climate-related security' appears to have found some purchase among ASEAN officials participating in an April 2022 webinar on climate change and security organised by the ASEAN Institute for Peace and Reconciliation (ASEAN-IPR) and the UN.

In the preparations for the meeting, particular sensitivity was exercised by the convenors to avoid the term 'climate security' to forestall any misunderstanding on the objectives of the meeting (which was largely to raise awareness of the risks of climate change to regional and global peace and security). Consequently, 'climate-related security' emerged as the preferred term during the meeting.[6] Important here is that the regional sensitivities regarding climate security lie less with the nature of the security threats themselves and more in the emphasis on which kinds of security threats to discuss. Hence, during the dialogue, most of the discussions on climate-related security issues dealt with food, water and other development issues rather than the potential for climate-induced conflicts.

The kinds of reservations found in regional discourses belie the fact that some ASEAN officials have engaged in UN climate security discussions. What is perceptible, though, is the nuanced stance on how to deal with climate challenges and the role of the UNSC. For example, in 2009 at the 63rd session of the UN General Assembly, then Indonesian Foreign Minister Marty Natalegawa remarked that:

> The security dimension of climate change should not be seen exclusively in terms of military conflict, endangering, and threatening international peace and security that requires the attention of the Security Council. Indonesia is cognisant that the impact of climate change is multi-dimensional and may include possible security implications ... Indonesia is of the view that the security dimensions of climate change relate principally to its development consequences ... at the heart of climate change is a core development challenge that carries potentially serious implications to development efforts. (Natalegawa 2009)

Indonesia's view on climate security and the UNSC has become more nuanced over the past decade, with its Foreign Minister Retno Marsudi acknowledging that 'the Security Council must consolidate efforts to better respond to the security impacts of climate change' (UNSC 2019). She also noted the need to 'better equip our peacekeepers with the capacity to undertake military operations other than war', including 'not only peacekeeping operations but also climate peace missions' (UNSC 2019). However, the bulk of Marsudi's statement nonetheless emphasised the importance of sustainable development in climate security and highlighted ASEAN's approach to climate security through humanitarian assistance and disaster response (UNSC 2019). Marsudi made the following distinction:

> our homework in the [Security] Council is to better define what falls under the ambit of climate change itself and what constitutes the security dimension of climate-related impacts ... For Indonesia, while the Security Council can deal with the security dimension of climate change, the United Nations Framework Convention on Climate Change remains the leading forum to address climate change. (UNSC 2019)

Statements from Malaysia, the Philippines and Viet Nam during the UNSC thematic debates on climate security were similarly nuanced. In 2015, the Malaysian representative noted that 'if left unchecked, climate change could in fact be the greatest threat multiplier for global security' and that it 'threatens SIDS [small island developing states] at the environmental, social and economic levels' (Ibrahim 2015). In 2021, the Philippines recognised that climate change exacerbated traditional drivers of conflict and stressed the 'importance of developing better risk-assessment and mitigation strategies for climate-related disasters' (UNSC 2021a).

Viet Nam saw the UNFCCC as the primary forum to discuss climate change, but acknowledged that the UNSC has a role and that 'all members should continue to discuss it in order to reach consensus, as the [UNSC] has demonstrated solidarity on many occasions' (UNSC 2021b). More recently, Viet Nam's Permanent Representative to the UN, Ambassador Dang Hoang Giang, expressed that while the UNSC needs to continue to promote the climate change and security agenda given its responsibility to maintain global peace and security, it 'should adopt a comprehensive approach to addressing the root causes of conflicts, striking a balance between traditional and non-traditional security challenges, and incorporating climate change risk analysis into the functions and mandates of peacekeeping and special political missions' (Vietnam News Agency 2023).

Given that ASEAN's comprehensive security approach is all-encompassing, tackling conflict issues like conflict prevention, peacekeeping or climate missions should be a non-issue. But, at the highest level, even as ASEAN leaders declare that climate change is an existential threat to national security requiring a comprehensive approach, their policy responses have focused on mitigation and adaptation.

Statements from the leaders of Indonesia, Malaysia and the Philippines, while highlighting the need for a comprehensive approach to climate change, place emphasis on addressing food security, economic security, environment security and water security. At the G20 Summit in 2021, Indonesian President Joko Widodo called on the G20 to 'set an example, ... to lead the world in cooperation on handling climate change and managing the environment in a sustainable manner by performing concrete actions' (Widodo 2021). Malaysia's Prime Minister Anwar Ibrahim has declared that 'our priority is to achieve low carbon and climate-resilient socio-economic development' (Othman 2023). Philippine President Ferdinand Marcos Jr stressed the need to 'work together to focus on strengthening food security, climate change mitigation and adaptation and developing the digital domain to encompass all levels of society in our countries to ensure connectivity and other cross-cutting issues' (Marcos 2023).

Cambodia's former Prime Minister Hun Sen, while declaring that the 'most prolonged danger we are facing remains climate change', appealed for concerted action to implement mitigation measures (Sochan 2023). Lao President Thongloun Sisoulith highlighted the need for increased resilience in the most vulnerable sectors and for accelerating decarbonisation (Sisoulith 2020). Meanwhile, Myanmar, Singapore, Thailand and Viet Nam have also drawn attention to the impact of climate change on water, energy and food security, underscoring the importance of sustainable development.

In sum, while most ASEAN leaders speak about climate change and security, the security concerns here refer to food, water, environment, energy, public health, and human security. By drawing more attention to the human security and developmental dimensions of climate change, and not using the term 'climate security', ASEAN states have avoided having to confront the issue of climate-induced conflicts and the compounded impact of climate on existing conflicts in Southeast Asia.

Teasing out ASEAN's reservations over 'climate security'

Reservations around the term 'climate security' stem from, first, the potential for interference in internal affairs when states are unable to provide adequate protection to their population in times of devastating disasters, particularly in situations that require international intervention.

The region's experience with the (mis)use of the doctrine of Responsibility to Protect (R2P) during the 2008 Cyclone Nargis in Myanmar set a bad precedent. Nargis had led to massive devastation of property and loss of lives; all told, around 2.5 million people were affected (APR2P 2008). Following the disaster, Myanmar's State Peace and Development Council (SPDC) refused to allow international aid into the country despite not being able itself to expeditiously provide assistance to affected communities. This led then French Foreign Minister Bernard Kouchner to call on the UNSC to invoke the R2P.

Although the R2P applies to collective action for mass atrocity crimes (genocide, war crimes, crimes against humanity, ethnic cleansing), the argument for invoking R2P on Myanmar was that the wilful neglect of the military junta when it comes to helping the victims and the deliberate attempt to withhold aid from people facing immediate risk of death were tantamount to crimes against humanity (APR2P 2021). France failed in its bid following objections from China and Russia and non-permanent members Indonesia and Viet Nam, but the episode left ASEAN states wary about what qualifies as timely and appropriate action by the international community to protect international peace and security (HLAP 2014, 9–32).

The second issue is the implications of climate security for regional stability and threats to international security. While studies have examined how climate change could compound existing conflicts in Iraq, Somalia and the Lake Chad region (CNA Corporation 2007; Vivekananda and Born 2018), experts in Southeast Asia contend that the region, unlike the Middle East and North Africa (MENA), is relatively stable even with the presence of internal conflicts in some states (Kassim 2014; Liu 2013; Suzuki 2021). Arguably, the kinds of conflict present in the region are not the ones that warrant the attention of the UNSC; of the 86 UNSC resolutions since 2006 dealing with a variety of R2P-related crises, none have applied to Southeast Asia or the wider Asian region.

The third reason is the capacity of regional mechanisms to deal with compounded climate-related risks. While ASEAN officials are quick to point out that their region is stable, internal conflicts exist in climate-sensitive areas like the southern parts of the Philippines, Southeast Papua in Indonesia and parts of Myanmar including the Rakhine region. Yet the ASEAN Coordinating Centre for Humanitarian Assistance on disaster management (AHA Centre) is not equipped to provide support in conflict settings. Further, the AHA Centre faces certain institutional constraints. Although its mandate is already confined to natural disasters and not man-made disasters (such as conflicts), it still has to have consent from an affected state for its operations. This has meant, for example, that despite the worsening humanitarian situation in Myanmar following the military coup in 2021, there is not much that the AHA Centre or ASEAN can do (Kamal 2023).

Last but not least is the wariness about climate security becoming a military-led agenda. The militaries in the region are usually the first responders in disasters, given the assets and organisational skills they can bring to search and rescue operations. However, the institutionalisation of ASEAN-led HADR exercises and the growing participation of militaries from outside the region have raised concerns that such activities could be used for geopolitical advantage (Trias and Cook 2023). It should not be discounted that external parties may be interested in expanding HADR exercises to include non-HADR related activities such as maritime patrols that have repercussions for maritime territorial disputes.

Moreover, territorial disputes related to fishing and sub-sea minerals in the South China Sea arguably could exacerbate the climate change effects in the area. These not only would escalate the simmering conflicts in the South China Sea but also could heighten major power rivalry between the United States and China, which would diminish opportunities for expanding the functional areas of cooperation among states to strengthen climate governance (Fetzek, Fleishman, and Rezzonico 2021).

Conclusion: the imperatives of climate security

In situating ASEAN's discourses as they relate to the security dimensions of climate change, this paper has shown that not only has the focus largely been on addressing the developmental consequences of climate change as part of the comprehensive framing of security in the region, but that discussion of climate-induced conflicts that threaten peace and security is not finding much traction in Southeast Asia and has been missing in regional deliberations.

As can be gleaned from the statements of ASEAN leaders and officials, the region's approaches to climate-related security risks lean more towards a comprehensive development approach involving multiple pathways to achieving security, by promoting sustainability and resilience-building, and strengthening regional cooperation on the environment, health and related sectors. ASEAN has also encouraged more institutionalised multilateral cooperation on HADR that is inclusive and aimed at fostering trust and building confidence among countries in the wider Indo-Pacific region. However, there is hardly any mention of developing early warning systems for conflict prevention and peacebuilding efforts.

ASEAN's preference for development-oriented approaches is compelling given the overwhelming challenges faced by affected countries, particularly in times of extreme weather events. The economic impact of disasters caused by extreme weather has been staggering, with annual losses estimated at USD 780 billion, with the figure projected to increase to between USD 1.4 trillion and 4.7 trillion by 2050 (McKinsey 2020). There are also the increasing threats to food security from lower production yields and to public health considering that environmental changes contribute to emerging infectious diseases.

However, the magnitude of the climate emergency extends well beyond the economy, food and health. The multiplicity of risks associated with climate change, such as scarcity of water and other resources, are also well-established drivers of conflict. As mentioned, there are countries in ASEAN that have extant domestic conflicts. The socio-economic impact of climate change, including the forced displacement of vulnerable groups and communities, could compound existing fragilities, which in turn could destabilise already vulnerable regions in Southeast Asia (Caballero-Anthony and Emmers 2022). These are more than sufficient reasons for the region to engage in climate-induced security threats. Certainly, it is doubtful how long ASEAN can continue to avoid confronting the issue given that the effects of climate change are already being felt, adversely affecting the well-being and security of its people.

Arguably, if ASEAN's notion of and approaches to security were indeed comprehensive, then political leaders and officials should be preparing the region to deal with possible instabilities arising from the consequences of forced displacements and competition over resources. To do this, there has to be discussion of these potential risks within the political and security circles in ASEAN, and within relevant sectors of the ASEAN Community's political-security pillar. By no longer confining discussions largely within its socio-cultural pillar, ASEAN and its relevant mechanisms could assess the kinds of regional capacities that need to be developed and strengthened to respond effectively to an array of security challenges from climate change. So, when ASEAN officials declare the need to build resilience, there should be discussions on how mechanisms like risk assessments and early warning systems for various threats, including conflict, can be established. Having these mechanisms would strengthen ASEAN's capacity to prevent conflicts and other destabilising issues that could threaten regional peace and security.

Similarly, while there are discussions among sectoral bodies within the ASEAN Community's socio-cultural pillar on how to develop anticipatory actions to mitigate the impact of severe weather events, there should also be discussions on how to develop a regional protection framework for climate-induced refugees and other forcibly displaced people, particularly in conflict-affected regions. Among the ideas mooted to advance a protection framework is to set up a regional monitoring system to track violence and human rights violations in certain settings (Leong and Ng 2023).

Moreover, while ASEAN mechanisms on conflict prevention and peacebuilding do not include establishing a regional peacekeeping force, existing institutions like ASEAN-IPR, the ASEAN Commission on the Promotion and Protection of the Rights of Women and Children and the ASEAN Intergovernmental Commission on Human Rights (AICHR) should be harnessed to initiate and facilitate regional dialogues among key stakeholders on how to respond to climate security challenges.

In other words, ASEAN's long-held notion of comprehensive security should be maximised to provide multiple pathways to achieve peace and security in a rapidly changing climate. In keeping with the regional maxim of 'no security without development, no peace without security', now is the time for climate security to have a place in ASEAN's three-pillared approach to a secure, stable and prosperous community. Advancing the agenda of climate security can go a long way in helping states and societies address climate-related security risks, promote closer regional cooperation and advance sustainable peace.

Disclosure statement

No potential conflict of interest was reported by the author.

Notes

1. Countries with a national Climate Emergency Declaration (CED) include Bangladesh, Japan, Singapore and South Korea. According to data from Cedamia (2020), the UK declared a national climate emergency on 1 May 2019, and the Republic of Ireland on 9 May 2019, followed by Canada, Argentina, Spain, Austria, France, Malta, Bangladesh, the European Union (in November 2019), Italy, Andorra, the Maldives, South Korea, Japan, New Zealand, Singapore, Fiji, then Peru and Vanuatu.
2. In this paper, Southeast Asia and ASEAN are used interchangeably.
3. Established in 1967, ASEAN comprises 10 states: Brunei, Cambodia, Indonesia, Laos, Malaysia, Myanmar, the Philippines, Singapore, Thailand and Viet Nam. Timor-Leste is officially an observer in ASEAN. In November 2022, ASEAN leaders agreed in principle to admit Timor-Leste into the bloc.
4. While tangentially linking climate change to environmental change, Homer-Dixon (1994) argues that environmental degradation could lead to environmental scarcity and could in turn cause the failure of state institutions, leading to high levels of intra-state violence.
5. Oels (2012, 185–205) observes that claims about climate change as a cause of violent conflict remain contested. Other scholars (Campbell and Parthemore 2008; Schwartz and Randall 2003) suggest that connecting climate to security could presage a military response and the logic of exceptionalism, and have instead proposed that desecuritising the climate issue would be a better response.
6. Personal communication with UN and ASEAN officials. The author was a participant in the dialogue which was co-organised and facilitated by an ASEAN body within the APSC pillar. Similar concerns were expressed at the 6th UN–ASEAN regional dialogue (AURED) held in Jakarta in November 2023 to discuss climate change and security, and at a follow-up meeting convened in February 2024 in Kuala Lumpur, Malaysia to propose including climate-related security issues in drafting the ASEAN APSC Post-2025 Blueprint.

ORCID

Mely Caballero-Anthony http://orcid.org/0000-0003-1064-745X

Bibliography

AEC (ASEAN Economic Community). 2015. *ASEAN Economic Community Blueprint 2025*. Jakarta: ASEAN Secretariat. https://asean.org/book/asean-economic-community-blueprint-2025/

Alagappa, M. 1988. "Comprehensive Security: Interpretations in ASEAN Countries." In *Asian Security Issues: Regional and Global*, edited by R. A. Scalapino, S. Sato, J. Wanandi, and S.-J. Han, 50–78. Berkeley, CA: University of California, Institute of East Asian Studies.

APR2P (Asia-Pacific Centre for the Responsibility to Protect). 2008. "Cyclone Nargis and the Responsibility to Protect." Myanmar/Burma Briefing no. 2. Brisbane: APR2P. https://r2pasiapacific.org/files/582/briefing_no2_cyclonenargis_r2p_myanmar.pdf

APR2P (Asia-Pacific Centre for the Responsibility to Protect). 2021. "Atrocity Crimes in Myanmar? Assessing Violence in the Wake of the 1 February 2021 Attempted Coup." Brisbane: APR2P. https://r2pasiapacific.org/article/2021/10/atrocity-crimes-myanmar

APSC (ASEAN Political-Security Community). n.d. "Maritime Cooperation." ASEAN Secretariat. Accessed May 25, 2024. https://asean.org/our-communities/asean-political-security-community/peaceful-secure-and-stable-region/maritime-cooperation/

ASCC (ASEAN Socio-Cultural Community). 2009. *ASEAN Socio-Cultural Community Blueprint 2015*. Jakarta: ASEAN Secretariat. https://asean.org/wp-content/uploads/images/archive/5187-19.pdf

ASCC (ASEAN Socio-Cultural Community). 2016. *ASEAN Socio-Cultural Community Blueprint 2025*. Jakarta: ASEAN Secretariat. https://www.asean.org/wp-content/uploads/2012/05/8.-March-2016-ASCC-Blueprint-2025.pdf

ASEAN (Association of Southeast Asian Nations). 2021. *ASEAN State of Climate Change Report*. Jakarta: ASEAN Secretariat. https://asean.org/wp-content/uploads/2021/10/ASCCR-e-publication-Correction_8-June.pdf

Australian Government, Department of Defence. 2009. *Defence White Paper*. https://www.defence.gov.au/about/strategic-planning/defence-white-paper

Barnett, J., R. A. Matthew, and K. L. O'Brien. 2010. "Global Environmental Change and Human Security: An Introduction." In *Global Environmental Change and Human Security*, edited by R. A. Matthew, J. Barnett, B. McDonald, and K. L. O'Brien, 3–32. Cambridge, MA: MIT Press.

Black, R., J. Busby, G. D. Dabelko, C. de Coning, H. Maalim, C. McAllister, M. Ndiloseh, D. Smith, et al. 2022. *Environment of Peace: Security in a New Era of Risk*. Stockholm: SIPRI. https://doi.org/10.55163/LCLS7037.

Boas, I., C. Farbotko, H. Adams, H. Sterly, S. Bush, K. van der Geest, H. Wiegel, et al. 2019. "Climate Migration Myths." *Nature Climate Change* 9 (12): 901–903. https://doi.org/10.1038/s41558-019-0633-3.

Boas, I., and D. Rothe. 2016. "From Conflict to Resilience? Explaining Recent Changes in Climate Security Discourse and Practice." *Environmental Politics* 25 (4): 613–632. https://doi.org/10.1080/09644016.2016.1160479.

Breckner, M., and U. Sunde. 2019. "Temperature Extremes, Global Warming, and Armed Conflict: New Insights from High Resolution Data." *World Development* 123 (3): 104624. https://doi.org/10.1016/j.worlddev.2019.104624.

Bretthauer, J. M. 2015. "Conditions for Peace and Conflict: Applying a Fuzzy-Set Qualitative Comparative Analysis to Cases of Resource Scarcity." *Journal of Conflict Resolution* 59 (4): 593–616. https://doi.org/10.1177/0022002713516841.

Brzoska, M., and C. Fröhlich. 2016. "Climate Change, Migration and Violent Conflict: Vulnerabilities, Pathways and Adaptation Strategies." *Migration and Development* 5 (2): 190–210. https://doi.org/10.1080/21632324.2015.1022973.

Busby, J. W. 2021. "Beyond Internal Conflict: The Emergent Practice of Climate Security." *Journal of Peace Research* 58 (1): 186–194. https://doi.org/10.1177/0022343320971019.

Buzan, B., O. Wæver, and J. de Wilde. 1998. *Security: A New Framework for Analysis*. Boulder, CO: Lynne Rienner.

Caballero-Anthony, M., and R. Emmers. 2022. "Keeping the Peace in Southeast Asia: ASEAN and the Quest for Positive Peace." *The Pacific Review* 35 (6): 1079–1104. https://doi.org/10.1080/09512748.2022.2075440.

Caballero-Anthony, M., R. Emmers, and A. Acharya, eds. 2006. *Non-Traditional Security in Asia: Dilemmas in Securitization*. London: Ashgate.

Caballero-Anthony, M., J. C. Trajano, A. D. B. Cook, S. Nanthini, J. M. L. Montesclaros, K. P. Landicho, and D. L. Goh. 2023. *Climate Change and Its Impact on Peace and Security in Southeast Asia*. Policy Report. Singapore: RSIS.

Campbell, K. M., and C. Parthemore. 2008. "National Security and Climate Change in Perspective." In *Climatic Cataclysm: The Foreign Policy and National Security Implications of Climate Change*, edited by K. M. Campbell, 1–25. Washington, DC: Brookings Institution Press.

Canyon, D. V., B. J. Ryan, and F. M. Burkle. 2017. "Military Provision of Humanitarian Assistance and Disaster Relief in Non-Conflict Crises." *Journal of Homeland Security and Emergency Management* 14 (3): 20170045. https://doi.org/10.1515/jhsem-2017-0045.

Cedamia. 2020. "Four Years of Climate Emergency Declarations." 5 December 2020. https://www.cedamia.org/climate-emergency-declarations/four-years-of-climate-emergency-declarations/

Climate Emergency Declaration. 2023. "Climate Emergency Declarations in 2,343 Jurisdictions and Local Governments Cover 1 Billion Citizens." 22 August 2023. http://climateemergencydeclaration.org/climate-emergency-declarations-cover-15-million-citizens/

CNA Corporation. 2007. *National Security and the Threat of Climate Change*. Alexandria: CNA Corporation https://www.cna.org/reports/2007/national-security-and-the-threat-of-climate-change

Cook, A. D. B., and S. Nanthini. 2023. *Mitigate, Adapt, Prepare: A Climate Security Agenda for the ASEAN Defence Ministers' Meeting*. Policy Report. Singapore: RSIS.

Day, A., and J. Caus. 2020. *Conflict Prevention in an Era of Climate Change: Adapting the UN to Climate-Security Risks*. New York: United Nations University.

Dellmuth, L. M., M.-T. Gustafsson, N. Bremberg, and M. Mobjörk. 2018. "Intergovernmental Organizations and Climate Security: Advancing the Research Agenda." *WIREs Climate Change* 9 (1): e496. https://doi.org/10.1002/wcc.496.

Eckersley, R. 2007. "Ecological Intervention: Prospects and Limits." *Ethics & International Affairs* 21 (3): 293–316. https://doi.org/10.1111/j.1747-7093.2007.00101.x.

Emmers, R., M. Caballero-Anthony, and A. Acharya, eds. 2006. *Studying Non-Traditional Security in Asia: Trends and Issues*. Singapore: Marshall Cavendish Academic.

ESCAP (United Nations Economic and Social Commission for Asia and the Pacific). 2022. *Asia-Pacific Disaster Report: Pathways to Adaptation and Resilience in South-East Asia: Subregional Report*. Bangkok: ESCAP.

EU-ASEAN Business Council. 2023. *Energy Transition in ASEAN*. Singapore: EU-ASEAN Business Council. https://www.eu-asean.eu/wp-content/uploads/2023/04/Energy-Transition-in-ASEAN-2023_5-April-2023.pdf

Fetzek, S., R. Fleishman, and A. Rezzonico. 2021. *Climate Security and the Strategic Energy Pathway in Southeast Asia: Part of the 'World Climate and Security Report 2020' Briefer Series*. Washington, DC: Center for Climate and Security (CCS).

Gemenne, F. 2011. "Why the Numbers Don't Add Up: A Review of Estimates and Predictions of People Displaced by Environmental Changes." *Global Environmental Change* 21 (Supp. 1): S41–S49. https://doi.org/10.1016/j.gloenvcha.2011.09.005.

Gleick, P. H. 1993. "Water and Conflict: Fresh Water Resources and International Security." *International Security* 18 (1): 79–112. https://doi.org/10.2307/2539033.

Haacke, J., and N. M. Morada, eds. 2010. *Cooperative Security in the Asia-Pacific: The ASEAN Regional Forum*. London: Routledge.

Hardt, J. N., and J. Scheffran. 2019. *Environmental Peacebuilding and Climate Change: Peace and Conflict Studies at the Edge of Transformation*. Policy Brief 68. Tokyo: Toda Peace Institute.

HLAP (High Level Advisory Panel on the Responsibility to Protect in Southeast Asia). 2014. *Mainstreaming the Responsibility to Protect in Southeast Asia: Pathway towards a Caring ASEAN Community*. Report presented at the United Nations, New York, 9 September 2014.

Hofmann, C.-A., and L. Hudson. 2009. "Military Responses to Natural Disasters: Last Resort or Inevitable Trend." *Humanitarian Exchange Magazine* 44: 29–30.

Homer-Dixon, T. F. 1994. "Environmental Scarcities and Violent Conflict: Evidence from Cases." *International Security* 19 (1): 5–40. https://doi.org/10.2307/2539147.

Hulme, M. 2011. *Why We Disagree about Climate Change*. Cambridge: Cambridge University Press.

Ibrahim, R. 2015. "Statement by H.E. Ambassador Ramlan Ibrahim, Permanent Representative of Malaysia to the United Nations." Presented at the Security Council Open Debate on "The Peace and Security Challenges Facing Small Island Developing States," New York, 30 July 2015. https://www.un.int/malaysia/sites/www.un.int/files/Malaysia/2015-Statements-Security-Council/2015-07-30_-_statement_open_debate_sids.pdf

IPCC (Intergovernmental Panel on Climate Change). 2007. *Climate Change 2007: Impacts, Adaptation and Vulnerability. Contribution of Working Group II to the Fourth Assessment Report of the IPCC.* Cambridge: Cambridge University Press. https://www.ipcc.ch/assessment-report/ar4/

IPCC (Intergovernmental Panel on Climate Change). 2023. *Climate Change 2023: Synthesis Report. Contribution of Working Groups I, II and III to the Sixth Assessment Report of the IPCC.* Geneva: IPCC. https://www.ipcc.ch/ar6-syr/

Kamal, A. 2023. "Beyond the 'Egosystem': A Case for Locally Led Humanitarian Resistance." The Humanitarian Leader 041. Burwood: Centre for Humanitarian Leadership. https://doi.org/10.21153/thl2023art1907.

Kassim, Y. R. 2014. *The Geopolitics of Intervention: Asia and the Responsibility to Protect*. Singapore: Springer.

Krampe, F., K. Tarif, and M. Mobjörk. 2020. *Pathways of Climate Insecurity: Guidance for Policymakers.* Stockholm: SIPRI.

Leong, W. K., and D. Ng. 2023. "UN Special Envoy Calls on ASEAN to Form Regional Protection Framework for Those Affected by Myanmar Crisis." *CNA*, 2 February 2023. https://www.channelnewsasia.com/asia/myanmar-asean-united-nations-special-envoy-noeleen-heyzer-3249386

Liu, Y. 2013. "On the Great Power Intervention in the Middle East Upheaval and Political Trend in the Middle East." *Journal of Middle Eastern and Islamic Studies (in Asia)* 7 (2): 1–34. https://doi.org/10.1080/19370679.2013.12023221.

Marcos, F., Jr. 2023. "PBBM Cites Need for ASEAN Plus Three to Ensure Food Security." Philippine Information Agency, 6 September 2023. https://pco.gov.ph/news_releases/pbbm-cites-need-for-asean-plus-three-to-ensure-food-security/#:~:text=%E2%80%9CTo%20move%20the%20APT%20(ASEAN,encompass%20all%20levels%20of%20society

Matthew, R. A. 2000. "The Environment as a National Security Issue." *Journal of Policy History* 12 (1): 101–122. https://doi.org/10.1353/jph.2000.0007.

Maystadt, J.-F., and O. Ecker. 2014. "Extreme Weather and Civil War: Does Drought Fuel Conflict in Somalia through Livestock Price Shocks?" *American Journal of Agricultural Economics* 96 (4): 1157–1182. https://doi.org/10.1093/ajae/aau010.

McDonald, M. 2013. "Discourses of Climate Security." *Political Geography* 33: 42–51. https://doi.org/10.1016/j.polgeo.2013.01.002.

McDonald, M. 2023. "Immovable Objects? Impediments to a UN Security Council Resolution on Climate Change." *International Affairs* 99 (4): 1635–1651. https://doi.org/10.1093/ia/iiad064.

McGuirk, E., and M. Burke. 2020. "The Economic Origins of Conflict in Africa." *Journal of Political Economy* 128 (10): 3940–3997. https://doi.org/10.1086/709993.

McKinsey. 2020. *Global Institute Report on Climate Risk and Response in Asia.* McKinsey Global Institute. https://www.mckinsey.com/capabilities/sustainability/our-insights/climate-risk-and-response-in-asia

Methmann, C., and A. Oels. 2015. "From 'Fearing' to 'Empowering' Climate Refugees: Governing Climate-Induced Migration in the Name of Resilience." *Security Dialogue* 46 (1): 51–68. https://doi.org/10.1177/0967010614552548.

Mosello, B., L. Rüttinger, and L. Sauerhammer. 2020. "The Climate Change-Conflict Connection." Berlin: CSEN. https://adelphi.de/en/publications/the-climate-change-conflict-connection

Myers, N. 1995. *Environmental Exodus: An Emergent Crisis in the Global Arena*. Washington, DC: The Climate Institute.

Natalegawa, M. 2009. "Climate Change and Its Possible Security Implications: Indonesia." https://www.un.org/esa/dsd/resources/res_pdfs/ga-64/cc-inputs/Indonesia_CCIS.pdf

Newby, V. 2019. "ANZUS Cooperation in Humanitarian Assistance and Disaster Response in the Asia-Pacific: Ships in the Night?" *Australian Journal of International Affairs* 74 (1): 72–88. https://doi.org/10.1080/10357718.2019.1693497.

Niazi, T. 2022. "Should Elevation of Climate Change to a 'National Security Threat' Be Scorned or Celebrated?" *Peace Review* 34 (1): 31–40. https://doi.org/10.1080/10402659.2022.2023427.

Oels, A. 2012. "From 'Securitization' of Climate Change to 'Climatization' of the Security Field: Comparing Three Theoretical Perspectives." In *Climate Change, Human Security and Violent Conflict*, edited by J. Scheffran, M. Brzoska, H. G. Brauch, P. M. Link, and J. Schilling, 185–205. Berlin: Springer. https://doi.org/10.1007/978-3-642-28626-1_9.

Othman, A. F. 2023. "ASEAN's Summits with Japan, South Korea Puts Focus on Green Technology." *New Straits Times*, 6 September 2023. https://www.nst.com.my/news/nation/2023/09/951887/aseans-summits-japan-south-korea-puts-focus-green-technology

Purvis, N., and J. Busby. 2004. *The Security Implications of Climate Change for the UN System*. Washington, DC: Woodrow Wilson Center for Scholars.

Rasmussen, K., and T. Birk. 2012. "Climate Change, Tipping Elements and Security." In *National Security and Human Health Implications of Climate Change*, edited by H. Fernando, Z. Klaić, and J. McCulley, 39–48. Dordrecht: Springer. https://doi.org/10.1007/978-94-007-2430-3_4.

Salehyan, I., and C. S. Hendrix. 2014. "Climate Shocks and Political Violence." *Global Environmental Change* 28: 239–250. https://doi.org/10.1016/j.gloenvcha.2014.07.007.

Scartozzi, C. M. 2022. "Climate Change in the UN Security Council: An Analysis of Discourses and Organizational Trends." *International Studies Perspectives* 23 (3): 290–312. https://doi.org/10.1093/isp/ekac003.

Schwartz, P., and D. Randall. 2003. *An Abrupt Climate Change Scenario and Its Implications for United States National Security*. Washington, DC: US Department of Defense.

Scott, S. V. 2008. "Securitizing Climate Change: International Legal Implications and Obstacles." *Cambridge Review of International Affairs* 21 (4): 603–619. https://doi.org/10.1080/09557570802452946.

Seah, S. 2022. "Asean." In *The Geopolitics of Climate and Security in the Indo-Pacific*, edited by R. Glasser, C. Johnstone, and A. Kapetas, 123–128. Barton: Australian Strategic Policy Institute.

Seah, S., and M. Martinus. 2021. *Gaps and Opportunities in ASEAN's Climate Governance*. Singapore: ISEAS-Yusof Ishak Institute.

Security Council Report. 2022. *The UN Security Council and Climate Change: Tracking the Agenda after the 2021 Veto*. New York: Security Council Report. https://www.securitycouncilreport.org/research-reports/the-un-security-council-and-climate-change-tracking-the-agenda-after-the-2021-veto.php

Sisoulith, T. 2020. "Statement by H.E. Mr. Thongloun Sisoulith, Prime Minister of Lao People's Democratic Republic." Permanent Mission of the Lao People's Democratic Republic to the United Nations, 13 December 2020. https://www.un.int/lao/statements_speeches/statement-he-mr-thongloun-sisoulith-prime-minister-lao-people%E2%80%99s-democratic-0

Sochan, R. 2023. "Climate Change Worse Threat than Ukraine War: Hun Sen." *Phnom Penh Post*, 30 March 2023. https://www.phnompenhpost.com/national/climate-change-worse-threat-ukraine-war-hun-sen

Stern, N. 2006. *Stern Review: The Economics of Climate Change*. United Kingdom: n.p.

Suzuki, S. 2021. "Intervention and State Strength: Association of Southeast Asian Nations and Economic Community of West African States Compared." *International Area Studies Review* 24 (1): 35–54. https://doi.org/10.1177/22338659211010296.

Tang, S. M., T. H. Hoang, A. S. Qian, G. Ong, and T. P. T. Pham. 2020. *The State of Southeast Asia: 2020 Survey Report*. Singapore: ISEAS-Yusof Ishak Institute.

Trias, A. P. L., and A. D. B. Cook. 2023. "Military Humanitarian and Disaster Governance Networks in Southeast Asia: Framework and Analysis." *Disasters* 47 (1): 205–241. https://doi.org/10.1111/disa.12542.

von Uexkull, N., and H. Buhaug. 2021. "Security Implications of Climate Change: A Decade of Scientific Progress." *Journal of Peace Research* 58 (1): 3–17. https://doi.org/10.1177/0022343320984210.

UN (United Nations). 2009. *Climate Change and Its Possible Security Implications: Report of the Secretary-General*. A/64/250. https://digitallibrary.un.org/record/667264?ln=en

UN (United Nations). 2022. "Secretary-General Warns of Climate Emergency, Calling Intergovernmental Panel's Report 'a File of Shame', while Saying Leaders 'Are Lying', Fuelling Flames." Press Release, 4 April 2022. https://press.un.org/en/2022/sgsm21228.doc.htm

UNDP (United Nations Development Programme). 1994. *Human Development Report: New Dimensions of Human Security*. New York: UNDP https://hdr.undp.org/content/human-development-report-1994

UNEP (United Nations Environment Programme). n.d.-a. "Climate Security Mechanism (CSM)." Accessed June 2, 2024. https://www.unep.org/topics/fresh-water/disasters-and-climate-change/climate-security-mechanism-csm

UNEP (United Nations Environment Programme). n.d.-b. "The Climate Emergency." Accessed 2 June 2024. https://www.unep.org/climate-emergency

UNFCCC (United Nations Framework Convention on Climate Change). n.d. "COP15." Accessed 2 June 2024. https://unfccc.int/event/cop-15

UNSC (United Nations Security Council). 2019. *Security Council 8451st Meeting*. S/PV.8451.

UNSC (United Nations Security Council). 2021a. "8923rd Meeting: People, Countries Impacted by Climate Change Also Vulnerable to Terrorist Recruitment, Violence, Speakers Tell Security Council in Open Debate." United Nations, 9 December 2021. https://press.un.org/en/2021/sc14728.doc.htm

UNSC (United Nations Security Council). 2021b. "8926th Meeting: Security Council Fails to Adopt Resolution Integrating Climate-related Security Risk into Conflict-Prevention Strategies." United Nations, 13 December 2021. https://press.un.org/en/2021/sc14732.doc.htm

Vietnam News Agency. 2023. "Vietnam's Commitments Reiterated at UNSC's Open Debate on Climate Change." 15 June 2023. https://en.vietnamplus.vn/vietnams-commitments-reiterated-at-unscs-open-debate-on-climate-change/254712.vnp

Vivekananda, J., and C. Born. 2018. *Lake Chad Region: Climate-Related Security Risk Assessment*. Stockholm: Expert Working Group on Climate-related Security Risks. https://adelphi.de/en/publications/lake-chad-region

Widodo, J. 2021. "President Jokowi: Indonesia Wants G20 to Set Example." Speech delivered at the *G20 Summit*. Cabinet Secretariat of the Republic of Indonesia, 31 October 2021. https://setkab.go.id/en/president-jokowi-indonesia-wants-g20-to-set-example-in-handling-climate-change/

Witmer, F. D. W., A. M. Linke, J. O'Loughlin, A. Gettelman, and A. Laing. 2017. "Subnational Violent Conflict Forecasts for Sub-Saharan Africa, 2015–65, Using Climate-Sensitive Models." *Journal of Peace Research* 54 (2): 175–192. https://doi.org/10.1177/0022343316682064.

WWF. 2023. "UN Climate Science Report Set to Highlight Catastrophic Consequences of Climate Crisis." 9 March 2023. https://wwf.panda.org/wwf_news/?8027966/UN-climate-science-report-set-to-highlight-how-slow-response-to-climate-crisis-risks-catastrophic-consequences

Climate change in the Asia-Pacific security architecture: the case of ASEAN

Alistair D. B. Cook

ABSTRACT

Human-induced climate change will likely cause more intense and frequent cyclones and typhoons, as noted in the Intergovernmental Panel on Climate Change (IPCC) report *Climate Change 2021: The Physical Science Basis*. Within the Asia-Pacific region, regional security discussions at the leadership level focus on the imminent threats to states and societies of more frequent and intense extreme weather events and sea-level rise as it is the world's most exposed region. This article examines recent developments and discourse regarding climate change in the Asia-Pacific security architecture using the Association of Southeast Asian Nations (ASEAN) as a case study. It finds increasing recognition of climate change and its implications for peace and security within the regional security architecture. Debates over how best to address it to overcome an atomised approach to climate change impacts on peace and security in the region remain unresolved, but recent pronouncements at the leadership level signal an emerging mindset shift in understanding climate change within the regional security discourse. However, the substantive discussions largely take place in hybrid forums, albeit with official recognition in more traditional platforms, which may undermine efforts for an inclusive regional discourse on the climate change impacts on peace and security in ASEAN.

Introduction

The United Nations Economic and Social Committee for Asia and the Pacific (UN ESCAP) Disaster Report 2022 highlights the projections from the Sixth Assessment Report of the Intergovernmental Panel on Climate Change (IPCC) in the region. Under all climate change scenarios, and compared worldwide, the Asia-Pacific will be the region most impacted by heavy precipitation, followed by agricultural drought, hot temperatures/heatwaves, and warming winds intensifying tropical cyclones (ESCAP 2022). The report calls for countries to accelerate achievement of the Sustainable Development Goals and the targets for the Sendai Framework for Disaster Risk Reduction through climate action, leveraging new technologies to enhance adaptation, risk-informed financial investments, and to build regional cooperation to support ecosystem adaptation. It draws attention to the emerging disaster–climate–health nexus as an important focus for these efforts, signalling the need for more holistic and cross-sectoral action. While progress is found in national and local disaster risk reduction strategies, deaths from disasters and a lack of commitments towards the UNFCCC show a regression in climate action overall and indicate a widening gap between adaptation and mitigation efforts in the Asia-Pacific (ESCAP 2022).

With the projection of an increase in the frequency and intensity of extreme weather events and sea-level rise already facing the Asia-Pacific, there is a need to reconsider climate action in the region. Through the commitment of all member states of the Association of Southeast Asian Nations (ASEAN) to the UNFCCC and the Paris Agreement, climate action is recognised as a critical element in promoting regional security for states and societies across the region. There is a gap, however, between these commitments and the implementation of climate action within the regional security architecture. At present these efforts remain atomised in different sectoral siloes involved in specific policy issue areas. There are indications of an emerging regional narrative on climate change and its impacts on peace and security that captures it within the dominant collective and comprehensive approach to security in the region. Through documentation analysis of ASEAN statements, declarations, action plans, vision and outlook documents, blueprints and communiques related to climate change from 2015 to 2023, this article investigates the status of climate change in ASEAN and how this is reflected in the official security discourse in the region.

The discussion on climate change and security has been directly discussed in the academic literature since the 1980s. The links between climate change and international security have taken on additional significance since the United Nations Security Council agreed to debates in 2007 and 2011. Since then, there have been efforts by states to give higher priority to climate change issues and contribute to global discussions to negotiate and implement the 1995 Kyoto Protocol. These debates have not yet initiated significant policy changes or actions at the global level and are contested by many states, particularly in the Global South. In the last five years, we have seen European countries like the Netherlands host the Planetary Security Conference in 2017 and the European Union host climate-related security events in 2018. In the same year we saw Sweden use its position as the Security Council president to hold a thematic debate on climate change and security, which was the third time for the Council but the first since 2011 (Conca 2019). This stemmed from the first 2007 United Nations Security Council debate that recognised the impacts climate change has on international peace and security and aimed to build frameworks to address the concurrent risks of political instability, insecurity and conflict that arise from the interaction of climate change with social, economic and political factors.

Through the securitsation of climate change, there have been concerted efforts to assess the impacts of climate change on national and international security by individual governments, the UN, and collective security arrangements, as well as the scientific, academic and think tank communities. However, even though the Asia-Pacific is the region most exposed to extreme weather events and sea-level rise, relatively little space is given in academic and policy circles to exploring how states and societies in the region understand and organise to tackle the peace and security impacts of climate change at the regional level.

Belmont and De Coning (2022) argue that instead of chasing the perennial and hard-to-prove question of whether climate change causes conflict, complexity theory would suggest that the IPCC should instead focus on more realistic goals such as exploring how climate change and societal security are interrelated in specific contexts and thereby generate a taxonomy of pathways through which climate change may influence societal security

(Belmont and De Coning 2022, 24). While this article argues the notions of multi-level and multi-scale policy interactions are necessary to address climate security issues, the debates at the global level would benefit from broadening participation from people in the Global South and engaging the understandings of how regional institutions offer an important avenue for cooperation on the peace and security impacts of climate change. This would counter what Floyd (2015) cautions against: that while institutional fragmentation of global climate security governance is not automatically problematic, the phenomenon of ideational fragmentation is highly disadvantageous to achieving climate security for people (Floyd 2015, 120). It would therefore benefit global climate security governance if it were to become more inclusive of states and regional security communities around the world and employ comprehensive and non-traditional notions of security.

This article will first examine the climate impacts in the region. It will then consider the place of climate change in ASEAN before turning to how these discussions are organised and the emerging discourse within the regional security community. The article then considers these implications for how climate security is understood and acted upon in the region. It finds that where regional security discussions link climate change and security, it comes from a different perspective, primarily viewed through the disaster management and emergency response framework. While there is a collective and comprehensive understanding of climate security within the regional security community, the institutional make-up of the regional organisation has led to an atomised approach to climate security governance to date. However, official pronouncements made at the leadership level in ASEAN and with external partners suggest a growing momentum to integrate climate change across the regional security architecture through collective and comprehensive notions of security, but this has not yet led to institutional change.

Climate impacts in the Asia-Pacific

The Asia-Pacific as a region has not met the 2020 benchmark to achieve the Sustainable Development Goals (SDGs) by 2030. The Asia-Pacific is the region most exposed to disasters. Seven of the top 10 countries identified by the Global Climate Risk Index (CRI) as most affected by extreme weather events and their impacts over the period 1999–2018 are in the Asia-Pacific (Remmits and Birkman 2020, 5). Dhaka, Manila, Bangkok, Yangon, Jakarta, Ho Chi Minh and Kolkata are all categorised as at 'extreme risk' on the climate change vulnerability index (ADB 2013). Coastal cities are increasingly exposed to flooding and agricultural land and water supplies are impacted by saltwater intrusion. There is likely a need for new sources and access to water for drinking, irrigation and transportation. Indeed, only 50% of Goal 11: sustainable cities and communities is measurable in the region, and under the current trajectory all targets will be missed by 2030. The exposure to climate change impacts has not resulted in substantive climate action at the local, national or regional level across the Asia-Pacific.

The *Asia and the Pacific SDG Progress Report 2021* highlights that the region has substantially regressed on Goal 13: Climate Action. The Asia-Pacific has made some progress towards protecting coastal areas but economic benefits from sustainable fisheries and ocean quality have declined (UN 2021). Documentation and progress remain difficult to track with the lack of available data as one important component needed to further understand the impacts of climate change. Reviews and progress reports made available in the region tend to be forward-looking documents rather than providing an assessment of the progress made to date. This inevitably obfuscates the ability to better determine areas for improvement in reaching the 2030 targets. However, an analysis of official documentation allows us to track the regional discourse and assess rhetorical changes that take place at the leadership level as an indication of prospective policy directions.

Twenty years ago, Schwartz and Randall argued that the inclusion of investment is necessary in predictive modelling for alternate scenarios, assessments to determine vulnerabilities like climate impacts on existing agricultural, water, and mineral resources; technical capacity; and social cohesion to inform adaptive responses and that exploration of local implications and technological advancements to address potential and lived scenarios was necessary (Schwartz and Randall 2003). Despite the progress made over the subsequent decades, it has not resulted in the necessary substantive climate action, and available financial resources remain inadequate.

Under the current trajectory, this will mean that more financial resources and available personnel will be used in disaster preparedness and response activities unless there are significant policy changes. The impacts of climate change will require more humanitarian assistance and disaster relief, stability operations and search and rescue missions. Militaries will face more frequent and concurrent commitments that stretch resources and potentially reduce readiness for other efforts (New Zealand Ministry of Defence 2018). This will increase pressure on disaster management and security institutions to develop a better understanding of specific climate-related risks in coastal cities across the Asia-Pacific (Remmits and Birkman 2020). Climate change adaptation is now more a reactive necessity than a proactive strategy to build more sustainable environments for emerging climate scenarios in the Asia-Pacific. Since 2004, countries in the region have taken an active interest in international coordination and capacity building for disaster response and prevention (Prayitno 2021, 16). However, given the status of progress on these components, it has focused more on disaster response than prevention actions.

The focus of this section highlights the direct climate impacts in the Asia-Pacific of more frequent and intense extreme weather events and sea-level rise. While these developments are monitored across the region, it is important to investigate how they are understood and whether this determination can inform how climate impacts are governed by regional security architecture in the Asia-Pacific. This article will investigate these developments within ASEAN as one important regional security community in the Asia-Pacific.

Climate change and security: a review

Within the Asia-Pacific there are different local conditions often understood in terms of proximity – how geographically close two or more countries are; capacity – the state of development of a given country; and legitimacy – the ways in which governments are in power and how this is maintained or challenged. These dynamics have shaped the ways in which relations between countries are governed. When considering regional cooperation these have taken on an important security dynamic that influences the ways in which countries address the individual and shared challenges they face (Caballero-Anthony and Emmers 2006). The regional level of interaction between countries provides an important avenue to understand the ways in which shared challenges like climate change are shaped and how they reflect the dominant mindset of political elites across the region. It is also important to consider the relations across multiple levels in what Foot and Goh (2019) refer to as the conjunctions of the regional–global and the unit–regional–global levels of analysis – the 'grey areas' where social formations meet and interact – through the conjunctions analytical framework (Foot and Goh 2019). As this article will demonstrate, climate change has become increasingly prominent in these regional discussions.

This section will review the state of the literature in understanding the intersection of climate change and security and the different approaches taken. It argues that climate change as a security concern has broadly evolved during this time through three theoretical generations of understanding: national security, human security, and contemporary approaches. While this generational theoretical approach offers an evolutionary understanding of climate change, the experience in Southeast Asia as illustrated through the discourse and security practice in the regional security architecture is one of collective, comprehensive, and non-traditional approaches to security, which bridges the notions of national and human security. Matt McDonald (2013) argues that the distinctions between the different generations of climate security articulated here are partial and imperfect, with boundaries between them far from fixed or wholly exclusive (McDonald 2013, 49). The prospects of contemporary approaches moving towards ecological security, planetary security or planetary health approaches, while convincing, have not yet gained the necessary traction within the regional security architecture. However, in the documentation analysis undertaken in this article, we find the pronouncements made at the ASEAN leadership level indicate a growing traction for these approaches within the dominant notions of security articulated in ASEAN. The documentation analysis illustrates that the practice of security is one that is negotiated through contestation and compromise and produces a blended form of multiple theoretical approaches. In this sense, it is important to consider the theoretical evolution of climate change and security alongside the security practice of linking climate change and security in different forums.

National security

Over the past 40 years, the connections between climate change and security have evolved through different theoretical generations of scholarship. As Matt McDonald (2013) argued, the 'first generation' of climate change and security scholarship was where global warming was a threat to global peace and security and warranted attention at the highest political level as a threat to human survivability (McDonald 2013, 43). Alongside this articulation of a link between climate change and security was the link to environmental security where competition over natural resources would result in conflict between states, and climate change would elevate this conflict risk. The Middle East and North Africa was identified as the region most susceptible to such resource competition that highlighted the potential for 'water wars' (Starr 1991). Within the Asia-Pacific, this connection was applied to the trans-boundary Brahmaputra River that flows through Bangladesh, Bhutan, China and India, and the Mekong River that runs through China, Myanmar, Laos, Cambodia, Viet Nam and Thailand as a source of conflict (Hauger 2012, 36). As the impact of climate change became increasingly recognised as a security threat and conflict risk, it was captured as a 'threat multiplier' that would increase the risk of conflict between these countries. Through this link between climate change and security, advocates argued for greater urgency, more funding and its incorporation in the high politics of security. Such a link activates the security sector and military to engage in overcoming this threat to peace and security and was articulated in UN Security Council debates in 2007 and 2011.

Since this connection was articulated, the security sector and militaries have engaged in climate action in several ways. The 'conflict multiplier' effect now includes additional scenarios like the climate change impacts of fish migration leading to conflict between fishing communities as they seek catch outside territorial waters, and the increasing competition in the Arctic circle between claimant states as polar ice caps melt (Goodman et al. 2021). While these dynamics of climate change have garnered the attention of the security sector and militaries, their roles and responsibilities remain contested and ultimately focused on indirect climate impacts.

Since the 2000s the security sector and militaries have engaged in more direct climate impact action such as addressing their contributions to global carbon emissions, more frequent humanitarian assistance and disaster relief missions, the threat of sea-level rise to military installations and the impacts on militaries in new climate scenarios. Most of these developments are led by the most advanced militaries in Europe and North America and are articulated in their defence white papers as part of whole-of-government commitments to mitigate and adapt to global climate change (DOD 2021a, 2021b, 2022; NATO 2021). Within the Asia-Pacific, some national militaries have recognised climate change in their defence white papers or other national security documents, but this is mostly focused on the increasing frequency and intensity of extreme weather events and their role in humanitarian assistance and disaster relief missions (Cook and Nanthini 2023).

Human security

The greatest challenge to the national security framework was articulated in the 1994 United Nations Development Programme Human Development Report that advocated for human security. The human security paradigm sought to reconceptualise security focused on individuals and communities rather than on the survivability of nation states. This is the second generation of climate security scholarship that responded to the first-generation focus on conflict to articulate a more cooperative and people-centred politics (McDonald 2013). While recognising the benefit that the security framework brought to climate change discussions, it moved the referent from the state to the individual and community level. This debate saw discussions over community empowerment and brought together the 'freedom from want' – a developmental discourse – that many states in the Asia-Pacific articulated and those in Europe and North America who advocated the 'freedom from fear' – a rights-based discourse – as mutually reinforcing components of human security (UNDP 1994). The human security framework highlights the impact of climate change on people's access to sustainable food, water and energy supplies, the exposure of communities to more frequent and intense extreme weather events, and the disproportionate impacts of these developments on low-income households.

Elliott (2012) argues that within the security literature the move from a politics of security to a politics of adaptation and resilience building would be read as a de-securitisation of climate change in the Asia Pacific. Reading this move instead as '*human* securitisation' has the potential to sustain the tactical attractions of the language of security in the urgent attention that this brings to a problem while also redirecting security policy to securing the lives, livelihoods and, wherever possible, the lands and homes of those in the region who are most vulnerable and most insecure because of the threats of climate change (Elliott 2012, 12). This approach seeks to rearticulate security focused at the individual level.

Collective, comprehensive and non-traditional security

The 1997 Asian Financial Crisis was an inflection point that brought countries in the Asia-Pacific some political and economic change. From this experience security scholars sought to bridge the divide between national and human security with a focus on the impacts of states and societies. This reconfiguration built on a comprehensive understanding of security that was evident in developing countries to raise their productivity and enhance human development. This commitment became recognised as non-traditional security concerns in national and regional security discussions across the Asia-Pacific in the 2000s onwards (Caballero-Anthony and Emmers 2006). Scholars from across the Asia-Pacific defined non-traditional security as issues that are challenges to the survival and well-being of peoples and states that arise primarily out of non-military sources, such as climate change, resource scarcity, infectious diseases, natural disasters, irregular migration, food shortages, people smuggling, drug trafficking and transnational crime. These dangers are often transnational in scope, defying unilateral remedies and requiring comprehensive – political, economic, social – responses, as well as the humanitarian use of military force (Caballero-Anthony 2016). Through bridging these two approaches it sought to navigate the political realities to provide an avenue to address security concerns at the local, national and regional levels.

Krampe and Mobjörk (2018) argue that the broader academic debate on climate-related security risks that emerged in the early 2000s involved a broad understanding of security that spans the divide between human and state security similar to this notion of non-traditional security that emanated from Southeast Asia, yet the global debate differed as it has to a large extent been shaped by traditional security concerns, and more specifically the potential linkages between climate change and violent conflicts (Krampe and Mobjörk 2018). Some scholars as recently as 2018 suggested that the increasing intensity, complexity, and frequency of security threats has caused governments, industry, non-governmental organisations, policymakers and communities throughout the Asia-Pacific to urgently reassess their exposure to security risks and vulnerabilities that has contributed to a transformation in our understanding and perception of security in the Asia-Pacific (Prizzia and Levy 2018, 51). However, regional scholarship shows that more collective, comprehensive and non-traditional security conceptions have been in use much longer than appreciated in global-level discussions.

Contemporary security approaches

As climate change impacts different sectors, the interlinkages between them become more visible. Since the 2000s and the prominence given to the Anthropocene in security discussions, fields of research have reconsidered their approaches to include human impacts on the natural environment and the interlinkages between the two. This has arguably led to the third generation of climate security scholarship. This was notably recognised in 2015 with the formation of the Rockefeller Foundation–Lancet Commission on Planetary Health. The commission builds on work in public health since the early 1980s that traditionally focused on treating human health without considering the surrounding natural ecosystems.

Planetary health includes the human impacts on the natural world and sees the earth system as interconnected.

While this narrative emerged out of the field of public health, a similar reconfiguration was underway within the field of security studies. Scholars argued for a shift away from using the environment to ecosystems as the referent. This framework recognises the inextricable links between the human and natural worlds rather than seeing the natural environment as something to be exploited or a source of threat. Given the actions of human beings are so pervasive, they have consequences which are complex and interrelated to the earth system (McDonald 2018, 165).

The concept of planetary health mirrors components of ecological security, and both are focused on cooperative politics. In 2015 the emergence of planetary security also recognised the human impacts on our complex and interrelated Earth system, as seen in the 2018 Hague Declaration on Planetary Security. Planetary security integrates the threat multiplier narrative into the discussion that focuses on the potential conflict hot spots that emerge from human activities and their consequences for the world around us. The third generation of climate security has begun to garner traction with decision makers and political elites. Yet while the emergence of complexity has nuanced our understanding of the world in which we live, it has not yet resulted in a significant shift in policy priorities and action.

There are concerns that climate security encourages a series of responses that reproduce inequalities through protecting relatively secure populations, namely wealthy and politically connected groups in both the Global North and Global South. Importantly, the criticism highlights that perceptions of and responses to vulnerabilities caused by climate change are built upon pre-existing systems of inequality (Thomas and Warner 2019, 9). While there is increasing recognition of the intersection of different ministries and sectors, it remains to be seen whether such integration is moving at sufficient scale and speed to meet climate realities. As Matt McDonald (2021) argues, while concepts like ecological security might be preferable, they remain marginal. Through an analysis of the statements made in ASEAN this article shows that while there is not an articulated climate security narrative, there is increasing recognition of climate change impacts on the regional security architecture.

Regional security arrangements

Across the Asia-Pacific, countries have developed regional economic and security arrangements. This section investigates the framing of climate change in the context of ASEAN. ASEAN is the most prominent avenue through which countries engage within the region and with their external partners on security matters. ASEAN is a notable example of facilitating wider participation in regional economy and security which engages the impacts of climate change and provides a mechanism for cooperation.

Tangney et al. (2021) observe in their systematic review that the literature is in agreement that maintaining and enhancing climate security in the Indo-Pacific will depend on (1) the region's capacity to strategically coordinate; (2) the willingness of governments to cooperate with communities and each other despite existing tensions; and (3) the government's ability to manage limited resources effectively and their capacity to identify and address climate-maladaptive path dependencies (Tangney et al. 2021, 167). The limitation to this study is that it does not consider ASEAN in its attempts to sample activity on climate security in the Indo-Pacific. More recently, Bremberg, Mobjörk, and Krampe (2022) note that Asian international organisations (IOs) focus on how climate change affects societies in their own geographic areas … and acknowledge the broader scope of climate-related security risks. They develop a framework on climate security that consists of three building blocks: discursive framing, institutional design and policy action. In the first instance the framework assesses whether a specialised office exists on climate security or if the IO has undertaken several institutional initiatives to enhance the work of the organisation, such as establishing a disaster management centre and working group on climate security. The third building block of the framework focuses on policy action to implement the climate security agenda through different areas that could include knowledge provision, risk assessment, collaboration, funding mechanisms and programming (Bremberg, Mobjörk, and Krampe 2022, 348–349). They observe that there are multiple frameworks aimed at supporting activities linked to specific climate-related security risks such as disaster management in ASEAN (Bremberg, Mobjörk, and Krampe 2022, 350).

Some scholars argue that emergency managers, security professionals and governments must promote climate adaptation and mitigation measures that protect communities in the Asia Pacific region (Prizzia and Levy 2018). However, it is the contention of this article that these practitioners are already doing so within their mandated areas but lack an overarching framework that connects their activities to those in other sectors and the broader ASEAN community, consisting of the three political-security, economic and socio-cultural community pillars. It is further argued that these efforts consist of several parts that are focused on specific climate-related security issues, as contended by Bremberg, Mobjörk, and Krampe (2022), but that recent pronouncements at the ASEAN leadership level indicate the inclusion of climate change in regional security discourse. While an atomised approach currently dominates the landscape, there is growing momentum to apply its regional security approach to climate change, as this article will now advance.

ASEAN

Since ASEAN was formed in 1967 by Indonesia, Malaysia, the Philippines, Singapore and Thailand, the member states of the regional organisation have built a system on consensus decision-making and non-interference. Membership grew to 10 member states with the addition of Brunei, Cambodia, Laos, Myanmar and Viet Nam. In 2008 ASEAN entered a new phase, with member states signing the ASEAN Charter that importantly committed them to a people-centred organisation to ensure peace, stability and prosperity, and to hold biannual ASEAN summits (ASEAN 2008). ASEAN leaders declared the 2009–2015 roadmap consisting of three community blueprints – political security, economic and socio-cultural – to form the basis for the ASEAN Community launched in 2015. Each community generates overarching plans of action and work plans for sectoral implementation.

Importantly, in 2008, member states signed the ASEAN Agreement on Disaster Management and Emergency Response (AADMER). AADMER formalised processes and led to the establishment of the ASEAN coordinating centre for Humanitarian Assistance on disaster management (AHA Centre) in 2011 to facilitate support to member states. Also in 2011, ASEAN member states established the ASEAN Working Group on Climate Change to implement the climate action and cross-sectoral work contained in the first ASEAN Socio-Cultural Blueprint (2009–2015) and subsequent five-year plans. The recent ASEAN Joint Statements further commit to

> *enhancing climate actions and cooperation* in ASEAN in accordance with the current global development of climate change related agenda, where and as appropriate, taking into account national circumstances and development priorities, including *through the ASEAN Working Group on Climate Change (AWGCC) Action Plan*. (ASEAN 2023a, 1, emphasis added by italics; see also ASEAN 2022)

However, as Seah and Martinus (2021) argue, the location of the AWGCC in the ASEAN Socio-Cultural Community lacks a mandate to coordinate beyond the ASEAN Ministerial Meeting on the Environment working groups. Given the cross-cutting nature of climate change, this has led to a proliferation of climate change-lined initiatives across different sectors in the regional organisation without an overarching framework (Seah and Martinus 2021, 14). The ASEAN Socio-Cultural Community (ASCC) Council meets biannually and oversees and coordinates all sectors within the ASCC. The ASCC Council Joint Statements from 2015 to 2023 only refer to climate change through the ASEAN Joint Statements on Climate Change to the UNFCCC. So it is in these joint statements that we can see how discussion of climate change is shaped within the regional security architecture. The 2023 ASEAN Joint Statement to the 2023 United Nations Climate Change Conference or Conference of the Parties of the UNFCCC, more commonly known as COP28 reaffirms commitment 'to strengthening ASEAN's *capabilities* to address climate-related disasters, reduce risks, and adapt to them by leveraging *primarily* existing mechanism *under the ASEAN Agreement on Disaster Management and Emergency Response (AADMER)*' (ASEAN 2023a, 1, emphasis added by italics.). This highlights the view of member states that climate-related disasters are a primary concern for ASEAN, and this is mandated first and foremost through AADMER. So, for the regional security community, climate security as discussed in various forums around the world has been understood within Southeast Asia primarily in relation to climate-related disasters as a preventive, disaster management and emergency response issue.

These commitments illustrate that at the regional (ASEAN) level collective capabilities are seen through different sectors (see Tables 1 and 2). Earlier research suggests that while there have been several initiatives like those outlined above, they are hampered because the ASEAN Political-Security Community has not been involved in the work on climate change, coordination between ASEAN Communities is insufficient, and there is a lack of commitment for member states to act (Krampe 2018). If we turn to the annual Joint ASEAN Defence Ministers Meeting Joint Statements, we can see the relative position of climate change for the defence sector, a key sector and traditional security actor within ASEAN under its Political-Security Community, indicating more recent changes.

After the establishment of the ASEAN Community in 2015, only since the Joint Declaration of the ASEAN Defence Ministers on Defence Cooperation to Strengthen Solidarity For a Harmonised Security (signed in Phnom Penh, Cambodia, on 22 June 2022) do the Joint Statements make explicit reference to climate change (ADMM 2022; see Table 1). It is again referenced in the subsequent Jakarta Joint Declaration of the ASEAN Defence Ministers Meeting for Peace, Prosperity and Security, signed on 25 November 2023 (ADMM 2023). Both declarations recognise the compounding impact that climate change has on the growing challenges posed by geopolitical and geostrategic shifts in the region. Perhaps more important is the shift in mindset to systemic security, where the Joint Statements since 2018 have articulated in broad terms the 'increasing complexity of nontraditional and transnational security challenges' (ADMM 2018, 1). This shift in mindset in the joint statements from 2018 to 2023 explicitly recognises the interconnected nature of security, including climate

Table 1. ASEAN official statements, declarations and communiques related to climate change 2015–2023.

Year/document	Relevant content to climate change 2015	2016	2017	2018	2019	2020	2021	2022	2023
ASEAN Joint Statement on Climate Change to the United Nations (UN) Framework Convention on Climate Change	Recognise the extreme vulnerability of ASEAN member states to climate change	Strengthen capabilities to deal with climate-related disasters through ASEAN Agreement on Disaster Management and Emergency Response (AADMER).	Strengthen capabilities to deal with climate-related disasters through AADMER, given the extreme vulnerability of ASEAN member states to climate change	Strengthen capabilities to deal with climate-related disasters through AADMER early warning system to reduce the consequences of climate-related disasters	Strengthen capabilities to deal with climate-related disasters through AADMER	N.A.	Strengthen capabilities to deal with climate-related disasters through AADMER Mainstream Climate Change Adaptation (CCA) and Disaster Risk Reduction (DRR).	Strengthen ASEAN's capabilities to deal with climate-related disasters and strengthen risk reduction measures through existing mechanisms under AADMER ASEAN Working Group on Climate Change	Strengthen capabilities to deal with climate-related disasters through AADMER Advance climate action through ASEAN Community-Based Climate Action Progress to establish ASEAN Centre for Climate Change Grave concern about rise in loss and damage associated with climate change, especially due to climate-related extreme weather events and slow onset disasters
ASEAN–UN Summit – Chairman's Statement.	7th Summit: Yes ASEAN–UN Work Plan on Environment and Climate Change (2016–2020) ASEAN–UN Strategic Plan of Action on Disaster Management	8th Summit: Yes ASEAN–UN Action Plan on Environment and Climate Change (2016–2020), which is aligned with the Declaration on ASEAN Post-2015 Environmental Sustainability and Climate Change Agenda	9th Summit: Yes ASEAN–UN Action Plan on Environment and Climate Change (2017–2020)	N.A.	10th Summit: Yes 'ASEAN–UN Environment and Climate Change Action Plan' (2016–2020), which has supported the implementation of the 'ASEAN Strategic Plan on Environment (ASPEN)' 'reduces vulnerability'	11th Summit: Enhanced cooperation under ASEAN–UN Plan of Action (2021–2025) on climate action Specifically through ASEAN–UN Environment and Climate Change Action Plan' (2016–2020)	N.A.	12th Summit: Yes follow up on recent dialogue activities by ASEAN Institute of Peace and Reconciliation IPR and the UN, on early warning mechanisms and climate-related security threats	13th Summit: Yes ASEAN Centre for Climate Change Enhance cooperation and broaden scope of collaboration through the 1st ASEAN–IPR–UN Track 1.5 Dialogue on Advancing Climate, Peace and Security
ASEAN Leaders' Declarations/ Chairman's Statements	Commit to resilience through preventing the generation of new risks and adapting to a changing climate through the implementation of economic, social, cultural, physical, and environmental measures which address exposure and vulnerability		Address climate change	Enhance civil–military coordination in ensuring ASEAN's collective and effective response to natural disasters and emergencies.	Concern for the multi-faceted impact and risks of climate change and climate-related disasters on the people of ASEAN	Recognises emerging challenges and the need to strengthen responses to adverse affects	Recognises the complex and cross-cutting challenges including conflict, climate change and others	Supports ASEAN Joint Declaration to United Nations Climate Change Conference or Conference of the Parties of the UNFCCC, more commonly known as COP	Global challenges are becoming more frequent, interconnected, and multidimensional, from the implications of climate change

(Continued)

Table 1. Continued.

Year/document	Relevant content to climate change								
	2015	2016	2017	2018	2019	2020	2021	2022	2023
ASEAN Foreign Ministers Joint Communique	increasing resilience to climate change, natural disasters and other shocks We noted with great concern that climate change is already having significant impact in the region, posing challenges to our environment, causing severe social and economic disruption and damage throughout the region. Notes ASEAN Action Plan on Joint Response to Climate Change for environmental sustainability Recognises Declaration on Institutionalising the Resilience of ASEAN and its Communities and People to Disaster and Climate Change Work with Canada to address regional security challenges, including climate change Identify comprehensive solutions to address regional challenges, including poverty eradication, disaster management and climate change Recognises ASEAN Joint Statement on Climate Change 2015 Recognises Declaration on ASEAN Post-2015 Environmental Sustainability and Climate Change Agenda Implement the ASEAN–US Strategic Partnership (2016–2020) on transnational challenges including climate change	Climate change has already caused major loss and damage throughout the ASEAN region, disproportionately affecting developing countries and impacting agriculture, energy supply, livelihoods, water security, land use and biodiversity Support development of ASEAN Joint Statement on Climate Change Implement ASEAN Action Plan on Joint Response to Climate Change for environmental sustainability Implement the ASEAN–US Strategic Partnership (2016–2020) including climate change Notes development of ASEAN–UN Work Plan on Environment and Climate Change 2016–2020	Gravely concerned about the multi-faceted impacts of climate change throughout the ASEAN region as outlined in the Assessment Reports (AR) of the IPCC and their disproportionate and adverse effects on vulnerable and marginalised groups,namely women, children, older persons, persons with disability, farmers, fisherfolk, persons in conflict-affected communities and those parts of basic sectors Reduce impact of 'climate-related disasters' Implement sustainable energy technology and early warning system to reduce impact of climate-related disasters Recognise ASEAN JC to UNFCC Recognises joint ASEAN–EU statement on Paris Agreement Strengthen ASEAN–Japan cooperation on climate change 'Promote ASEAN–UN complementarities to identify comprehensive solutions … to climate change.'	Recognise SAMCA Recognise ASEAN JC to UNFCC	'staying resilient and responsive to emerging challenges, including environmental and climate change issues and non-traditional security threats' Concern for the multi-faceted impact and risks of climate change and climate-related disasters on the people of ASEAN Highlighted the importance of addressing transnational challenges including climate change and advancing the 2030 Agenda for Sustainable Development, and looked forward to the 10th ASEAN–UN Summit in November this year	'multi-faceted impact and risks of climate change and climate-related disasters on the people of ASEAN' Cross-cutting issue for sustainable development' Recognise ASEAN JC to UNFCC Staying resilient and responsive to emerging challenges, including environmental and climate change issues and non-traditional security threats	Recognise ASEAN JC to UNFCC	Establish ASEAN Centre for Climate Change (ACCC) Welcomed Glasgow Climate Pact	Recognised development of ASEAN Joint Communique to UNFCC. Welcomed finalisation of ACCC 'Promote nature-based solutions and ecosystem-based approach to enhance the region's resilience … addressing climate change …' Reaffirmed the importance of implementing the Bandar Seri Begawan Declaration on the Strategic and Holistic Initiative to Link ASEAN Responses to Emergencies and Disaster (ASEAN SHIELD) to prepare for and manage climate-related and other disaster risks and their impacts in the region through a holistic, cross-pillar and cross-sectoral approach

(Continued)

Table 1. Continued.

Year/document	Relevant content to climate change								
	2015	2016	2017	2018	2019	2020	2021	2022	2023
ASCC Summit	ASEAN Joint Statement to COP	ASEAN Joint Statement to COP	ASEAN Joint Statement to COP	ASEAN Joint Statement to COP	ASEAN Joint Statement to COP	Climate change, no specifics	ASEAN Centre for Climate Change ASEAN Joint Statement to COP ASEAN State of Climate Change Report		
ADMM Joint Declaration		No	No	No	No	No	No	Yes	Yes

Source: Author's elaboration.

Table 2. ASEAN periodic documents related to climate change, 2015–2023.

Document	Date	Content relevant to climate change
ASEAN Leaders' Declaration on Sustainable Resilience	2023	– the escalating uncertainty and complexity of the disaster risk landscape in the region and interconnected transboundary risks and threats due to multi-layered compounding natural hazards, increasing systemic risk, and the interrelated challenges of risks posed by growing population, rapid urbanisation, impacts of climate change, and unsustainable utilisation of land resources, resulting in negative environmental, social, and economic consequences and other crises, such as persistent impact from pandemics, air and plastics pollution, soil degradation, and global warming, and the need to strengthen the resiliency of the ASEAN Community – Emphasise commitment to disaster resilience through AADMER and work plans and ASEAN Disaster Resilience Outlook
ASEAN Outlook on the Indo-Pacific	2021	– Climate Change and Disaster Risk Reduction and Management as a possible area of cooperation
ASEAN–UN Environment and Climate Change Action Plan	2021 – 2025	– The continued convening of regular ASEAN–UN Track 1.5 workshops and regional dialogues on political-security cooperation on … the intersection with humanitarian action in disaster and conflict settings, and discussion on the potential linkages between climate change and conflict
Special ASEAN Ministerial Meeting on Climate Action (SAMCA) and Expanded-SAMCA (E-SAMCA) Chair's Summary	2018	– AMS noted that stewardship for climate change was being driven at the highest political levels – by ASEAN heads of state and governments – Strengthen ASEAN capability and capacity…to share information and best practices
ASEAN–EU Statement on the Paris Agreement	2017	– Work together with all stakeholders to combat climate change and its impacts
ASCC Blueprint 2025	2016	– Adapt and respond to social and economic vulnerabilities in times of climate change-related crises, disasters as well as emerging threats and challenges (5.4) – Establish platforms to empower people living in at-risk areas to become resilient by reducing their exposure and vulnerability to climate-related extreme events and other economic, social and environmental shocks and disasters (D.4.ii)
APSC Blueprint 2025	2016	– Enhance the role of the ASEAN Secretary-General including assisting the ASEAN Chair in carrying duties in emergency humanitarian situations and implement the work of ADMM Plus EWG on HADR – Implement AADMER as common platform for region – Promote synergy with ASEAN on disaster management through ACDM – No direct mention of climate change
ASEAN Community Vision 2025	2015	– A resilient community with enhanced capacity and capability to adapt and respond to social and economic vulnerabilities, disasters, climate change as well as emerging threats and challenges (12.4)
ASEAN Joint Statement on Climate Change	2015	– Minimising the risks of extreme weather events and other climate-driven disasters

(*Continued*)

Table 2. Continued.

Document	Date	Content relevant to climate change
Declaration on Institutionalising the Resilience of ASEAN and its Communities and People to Disaster and Climate Change	2015	– Exposure of ASEAN's peoples and communities to various natural hazards is rapidly rising, vulnerability to extreme weather events and climate change is spiralling, and this 'new normal' leads to growing frequency and severity of disasters, exacting a tremendous toll on economic, social, cultural, physical, and environmental capital of countries and communities, which in turn stifles our region's sustainable development including our efforts towards poverty eradication – Urgent need to enhance awareness among member states and all ASEAN sectoral bodies and stakeholders to be able to simultaneously address the interlocking issues of disaster risk reduction, climate change adaptation and sustainable development and to effectively reduce adverse impacts of natural and human-induced disasters and climate change on key development sectors – Imperative to promote development that is transformative, inclusive, adaptive and sustainable, where peoples and communities thrive in peace and prosperity as envisioned in the ASEAN Community's Post-2015 Vision and its attendant documents and to build an ASEAN Community that is resilient to disasters and climate change, viewing resilience as a unifying but multifaceted process and outcome – Need for cross-pillar and cross-sectoral collaboration on disaster risk management, climate change adaptation and sustainable development and related cross-cutting concerns – Encourage all stakeholders to participate in planning and implementation of the institutionalisation of disaster risk management and climate change adaptation at the local, national and regional levels and monitor the progress and outcomes in terms of reducing disaster risk and adapting to climate change through multi-stakeholder means and mechanisms – Address underlying risk drivers and compounding factors, such as climate change and climate variability, uncontrolled urbanisation, ecosystem degradation, weak governance, limited risk management capacity especially at the local scale, poor management of urban and rural development, consequences of poverty and inequality, and conflict situations – Agree to enhance cooperation and collaboration among ASEAN member states to support the implementation of this declaration and develop a cross-pillar and cross-sectoral collaborative ASEAN work programme for resilience building in the context of disaster risk management, climate change adaptation and sustainable development – Assign the ASEAN Committee on Disaster Management (ACDM) as the focal point for cross-sectoral cooperation on resilience building at regional level, with the support of other relevant sectoral bodies

Source: Author's elaboration.

change. It also highlights the military as an emergent actor in climate security (Jayaram and Brisbois 2021). This shift is further underlined in the 2023 ASEAN Leaders' Declaration on Sustainable Resilience (ASEAN 2023c, 1) which recognises

> the escalating uncertainty and complexity of the disaster risk landscape in the region and interconnected transboundary risks and threats due to multi-layered compounding natural hazards, increasing systemic risk, and the interrelated challenges of risks posed by growing population, rapid urbanisation, and impacts of climate change ....

These are, in turn, consistent with the preventive, collective and comprehensive security approach that ASEAN has organised its affairs around since its founding.

The *ASEAN State of Climate Change Report* in 2021 also recognised this through its calls for the ASEAN community 'to develop a holistic ASEAN regional narrative to realise more transparent and ambitious climate action' (ASEAN 2021b, 1). While the ASEAN Charter committed the organisation to be larger than the sum of its parts (member states), this remains unrealised despite the launch of the ASEAN Community in 2015 – although the document analysis shows there is an emerging shift and recognition of the climate impacts on peace and security in the region.

One place there is increased visibility for the impacts of climate change on regional security architecture is through the UN–ASEAN Summit (see Table 1). The 2023 ASEAN Chairman's Statement on the summit drew attention to the strong cooperation on peace and security through continued engagement between the UN and the ASEAN Institute for Peace and Reconciliation (IPR). It recognised that this linkage would develop from its first ASEAN–IPR–UN Track 1.5 Regional Dialogue on Advancing Climate, Peace and Security Agenda in ASEAN in November 2023 (ASEAN 2023b). At the November dialogue, an independent report on *Climate Change and its Impact on Peace and Security in Southeast Asia* (Caballero-Anthony et al. 2023) was launched as part of the wider sixth ASEAN–UN Regional Dialogue on political-security cooperation. The report proposed a set of recommendations to integrate climate change impacts and effects through several policy entry points in ASEAN across its socio-cultural, economic, and political-security communities.

These recommendations indicate potential pathways for and understanding of the intersection of climate, peace and security in ASEAN. They include adopting policy frameworks to simultaneously address climate change and pursue conflict prevention and resolution, undertaking impact assessments, implement potential guidelines on digitalisation of food security, improving ASEAN institutional mandates, establishing a regional scientific platform, strengthening sub-national entities and cross-sectoral synergies, strengthening gender mainstreaming, providing a protection framework for displaced populations, and promoting closer UN–ASEAN cooperation on advancing a climate, peace and security agenda (Caballero-Anthony, et al. 2023, 86–90). This recognition at the ASEAN–UN Regional Dialogue highlights the status of climate security in ASEAN and recognises the areas for improvement. It finds that 'the compounding impact of climate change – interacting with various intractable peace and security challenges in the region – on vulnerable sectors and peoples could generate new security challenges to the well-being of states and communities, but also foster greater cooperation' (81). It highlights 11 key areas within the climate, peace and security realm within ASEAN: economic security, food security, health security, water security, natural resources, disasters, gender, forced migration, internal conflicts, and geo-political dynamics in the Mekong sub-region and in the South China Sea (Caballero-Anthony et al. 2023, 81–85).

The official statements at the UNFCCC and ASEAN–UN Summit further highlight climate-related natural hazards and sea-level rise as the main entry point for countries in the region to engage the global climate change agenda (see Table 1). Karki et al. (2023) argue that while existing security forums in the Asia-Pacific such as the ASEAN Defense Ministers Meeting Plus provide good foundations for intra-regional cooperation, the issue of climate security has thus far not featured prominently on the agenda (Karki et al. 2023, 14). The recent inclusion of the track 1.5 dialogue on climate, peace and security as part of the ASEAN–IPR–UN Regional Dialogue in 2023 and the references to climate change and systemic approaches in the ASEAN Defence Ministers' Meeting (ADMM) Joint Statements in 2022 and 2023 illustrate the potential beginning of a more regular and sustained effort for ASEAN to articulate its perspectives and understanding on these issues. The climate security discourse in ASEAN and related arrangements draws linkages to areas of successful cooperation to date alongside the areas for improvement. While there is presently limited engagement of the ASEAN Political-Security Community in climate security, recent developments signal growing recognition of the importance of climate change as part of a comprehensive agenda in the regional security community.

Unpacking climate security in the Asia-Pacific

Much of the literature on climate and security relates to whether climate change contributes to conflict and focuses on Europe, North America, and Australia (Krampe and Mobjork 2018). Less attention is paid to the understandings and developments of regional discourse in Southeast Asia and the wider Asia-Pacific security architecture. The International Military Council on Climate and Security Report on Climate Security and the Energy Security Pathway highlights a dominant trend in the literature that climate security is often focused on vulnerable populations in 'fragile' states, with less attention paid to sharpened competition over food, water, and energy resources between neighbouring states (Shidore, Fetzek, and Fleishman 2021). That said, there is some recognition of the potential for climate change to result in low-level conflict.

In a track 2/track 1.5 meeting in 2019, a Malaysian participant went beyond the Humanitarian Assistance and Disaster Relief (HADR) elements of climate change impacts on security to recognise that

> the prevalence of illegal, unreported and unregulated (IUU) fishing in our waters is partly due to global warming and it presents a threat to regional security that can develop into low-level inter-state conflicts. It is imperative for the AMS [ASEAN Member States] to work together to rein in IUU fishing and mitigate the problem. (Ahmad in NADI 2019, 7–8)

Such observations illustrate a nuanced approach recognising the multifaceted and complex interactive effects of climate change on regional security. It also illustrates what Foot and Goh (2019) identify as a 'grey area' conjunction where the regional security discourse operates in hybrid forums alongside the traditional track 1 level meetings of government-to-government interactions.

Furthermore, articulations captured in the document analysis draw attention to the location of discourse on the climate change impacts on regional security taking place and new avenues for cooperation that are recognised at the official level. This is seen in the statements made in official ASEAN documents (Tables 1 and 2) that recognise the work undertaken in more specialised forums. Ultimately, if a climate security agenda is meant to prioritise climate change impacts on security in the region, trust-building and cooperation will need to further incorporate climate change and disaster impacts across sectors. At present, while broad commitments to whole-of-society approaches are articulated in ASEAN (APSC Blueprint 2016), there remain challenges between individual ministries and agencies in the implementation of this vision. The locations of these discussions raise important questions over their efficacy and whether they signify greater specialisation and a more siloed approach placed farther from the people they are meant to engage (Hameiri and Jones, 2013) or if they offer new connections bringing disparate audiences together to address the issue of climate change impacts on regional peace and security.

While there are multiple ways that climate change impacts security and understandings of security, the loss of life from extreme weather events is a core security concern that is understandably dominant in regional discussions in the Asia-Pacific and in the IPCC Fifth Assessment Report chapter on human security (Adger et al. 2014 in Busby et al. 2018). The IPCC Fifth Assessment Report notes that while climate change may directly contribute to conflict, this link remains contested. The report nuances this with the indirect effects climate change has on conflict settings by its impact on per capita gross domestic product (GDP), economic contraction and weak state institutions that are strongly associated with violent conflict (Adger et al. 2014 in Busby et al. 2018). The challenge for the indirect link between climate change and conflict is that under this construction many parts of the Asia-Pacific could be placed within this category, and so this understanding of climate security faces rejection by the national governments and regional security communities as a back door to political interference in the domestic politics of member states, which runs in opposition to the non-interference norm in ASEAN. This may also provide a partial explanation of why these discussions on climate change impacts on regional peace and security are located in hybrid or 'grey area' conjunctions at arm's length from the official government level.

Through the official document analysis, the place of climate change within the discourse of traditional security actors references disasters as direct security concerns to countries in the Asia-Pacific, as the affected populations are significant in number and comparable to the human consequences of conflict, and they frequently require national and external military mobilisation in response (Busby et al. 2018, 90; ADMM Statements in Table 1). It is therefore important to understand that, rather than security being seen as the meeting of thresholds, it is the activation of military mobilisation as a component where civilian capabilities alone are insufficient to meet the needs of affected communities. Central to this understanding of regional security is that it is articulated as people-centred, and demand- rather than supply-driven.

However, the framing and approach of traditional security actor support for local systems is often put in stark terms, where the responsibility is on local institutions to access available funds rather than being engaged earlier on to shape the planning process and support these traditional security actors. Busby illustrates the former point by arguing that 'there is a strong record of targeted international support to strengthen state capacity for specific purposes, like investment in cyclone early warning systems and shelters. This requires local actors with vision and political space to take advantage of international assistance' (Busby 2021, 191). While the development of early warning systems has been advocated in multiple forums to increase the efficiency and effectiveness of humanitarian assistance and disaster relief, the availability of scientific data and technical know-how suffers from the lack of political will by government institutions to turn early warning into early action. This reinforces the need for greater engagement of states and societies earlier on in the decision-making and development process – a rearticulation of preventive notions of security. It is difficult to design effective climate action that provides solutions as a fait accompli to those who have little knowledge or buy-in to the finished policy. This ultimately is a challenge for the new avenues for cooperation in the hybrid or 'grey areas'.

ASEAN has invested in new avenues through which dialogue partners can engage on terms designed by those they seek to help. An example of this is the ASEAN Defence Ministers Meeting (ADMM)-Plus Expert Working Group (EWG) observer programme that allows non-Plus countries to observe selected EWG activities, and to share their expertise and experience to raise ASEAN member states' capabilities to address regional security challenges (MINDEF 2021). Other operational initiatives under consideration offer ways to support affected countries in the Asia-Pacific, such as the ASEAN and ADMM proposal to study the possibility of establishing a 'Regional Crisis Management and Response Centre' to coordinate HADR efforts, including information sharing with the proposed national coordination centres. It further considers 'developing more HADR scenarios in tabletop/field exercises to enhance civil–military cooperation and holding workshops under ADMM Plus EWG HADR that involve relevant stakeholders including civilian related agencies. All this would provide opportunity to share knowledge and experiences as well as improved coordination such as aligning/streamlining SOPs' (NADI 2019, 3).

One of the core difficulties that appears in how regional security communities engage with climate security issues is the convergence of strategic, operational and tactical components. While there appears to be broad agreement that climate change impacts the security of states and societies in ASEAN, less attention is spent on what to do about it and who should do it. Many of these discussions result in turning humanitarian assistance and disaster relief into the operational reality for different sectors for climate security concerns. As illustrated earlier, the AHA Centre was established over a decade ago as an operational entity to support ASEAN member states in disaster management and emergency response. The AHA Centre is located within the ASEAN Socio-Cultural Community that engages the disaster management sector, whereas recent suggestions that percolate around the region to engage militaries are located within the ASEAN Political Security Community. While these two communities, alongside the ASEAN Economic Community, are envisioned to work together, this has not yet occurred substantively on climate change. The regional organisation has yet to implement adequate supporting mechanisms for the climate realities they face. While ASEAN has made progress in this regard it remains in sectoral siloes and lacks a regional narrative that binds its institutions together for the climate realities of today – let alone the future. However, through the document analysis it is observed that there is increasing traction of climate impacts on security at the ASEAN–UN Summit, ASEAN engagement in the UNFCC, and ADMM levels.

Conclusion

From this investigation into the traction and understanding of climate change impacts on peace and security in ASEAN, we can see there is a common understanding that climate change has security implications for states and societies. The other components advocated as security threats within global debates have not led to a rearticulation or recategorisation of climate impacts as sources of conflict or expanded the security or defence sectors into new areas outside of humanitarian assistance and disaster relief or addressing institutional impacts of climate change. Rather, recognition of climate change impacts on peace and security remains focused on collective and comprehensive articulations within the regional security community. While the inclusion of climate change in security discussions has this particular focus, there are climate action efforts underway within individual line ministries. However, these developments remain disconnected from one another and pose a significant challenge for the ASEAN region.

ASEAN has been the driver of regional security governance and institution-building and is recognised as one of the more successful regional organisations. While it relies on informality and hybrid spaces to address regional security concerns, the pressure to recalibrate to deal with today's problems has become more urgent than ever (Caballero-Anthony 2022). Hence, there is a need to move towards a framework that transcends individual ministries, mandates and hybrid spaces in ASEAN and enables the region to find common cause and focus on the disaggregated climate impacts felt by states and societies. This will require not only cross-sectoral efforts but also increased vertical alignment in response to climate security threats, building synergies at the local, national, regional, and international levels (Thomas 2022, 75). It is unclear whether this concern is addressed with the activity undertaken in hybrid forums. However, the pronouncements at the ASEAN leadership level indicate a shift in the regional narrative to incorporate the climate impacts on security (see Tables 1 and 2). Despite these pronouncements and its signatory status to the UNFCCC and Paris Agreement, climate action in ASEAN presently remains disconnected, without a clear overarching narrative that would bridge the divides between the different hybrid forums and sectoral discussions. While there are broad commitments and initial indications for a more inclusive security discourse incorporating climate change, the discussion remains an ad hoc and embryonic element in the regional security architecture.

Disclosure statement

The author reports there are no competing interests to declare.

ORCID

Alistair D. B. Cook http://orcid.org/0000-0003-1934-5474

Bibliography

ASEAN Defence Ministers Meeting (ADMM). 2022. "Joint Declaration of the ASEAN Defence Ministers on Defence Cooperation to Strengthen Solidarity for a Harmonised Security." Phnom Penh, Cambodia, June 22. https://admm.asean.org/dmdocuments/2022_Jun_16th%20ADMM_Phnom%20Penh_22%20June%202022_8.%20Phnom%20Penh%20Vision%20on%20the%20Role%20of%20Defence%20Establishments%20in%20Support%20of%20COVID-19%20Recovery.pdf

Adger, W. N., Pulhin, J. M. Barnett, J. Dabelko, G. D. O. Spring, U. Vogel, C. H., et al. 2014. "Human Security." In Climate Change 2014: *Impacts, Adaptation, and Vulnerability. Part A: Global and Sectoral Aspects. Contribution of Working Group ii to the Fifth Assessment Report of the Intergovernmental Panel on Climate Change*, edited by C. B. Field, V. R. Barros, D. J. Dokken, K. J. Mach, M. D. Mastrandrea, T. E. Bilir, M. Chatterjee, 755–791. Cambridge, UK and New York, NY, USA: Cambridge University Press.

ADMM. 2018. "Joint Declaration of the ASEAN Defence Ministers on Strengthening Cooperation, Building Resilience." Singapore, October 19. https://www.nas.gov.sg/archivesonline/data/pdfdoc/MINDEF_20181019002/MINDEF_20181019001_1.pdf

ADMM. 2023. "Jakarta Joint Declaration of the ASEAN Defence Ministers Meeting for Peace, Prosperity and Security." Jakarta, Indonesia, November 25. https://admm.asean.org/dmdocuments/2023_Nov_17th%20ADMM_Jakarta_15%20November%202023_1.%20Joint%20Declaration.pdf

ASEAN. 2020. "ASEAN Outlook on the Indo-Pacific." Jakarta: ASEAN Secretariat. https://asean.org/asean2020/wp-content/uploads/2021/01/ASEAN-Outlook-on-the-Indo-Pacific_FINAL_22062019.pdf

ASEAN. 2021b. "ASEAN State of Climate Change Report: Current Status and Outlook of the ASEAN Region Toward the ASEAN Climate Vision 2050." Jakarta: ASEAN Secretariat. https://asean.org/wp-content/uploads/2021/10/ASCCR-e-publication-Correction_8-June.pdf

ASEAN. 2022. ASEAN Joint Statement on Climate Change to the 27th Session of the Conference of the Parties to the United Nations Framework Convention on Climate Change (UNFCCC COP-27). Jakarta: ASEAN. https://asean.org/wp-content/uploads/2022/11/33-ASEAN-Joint-Statement-on-Climate-Change-to-UNFCCC-COP-27.pdf.

ASEAN. 2023a. "ASEAN Joint Statement on Climate Change to the 28th Session of the Conference of the Parties to the United Nations Framework Convention on Climate Change (UNFCCC COP-28)." https://asean.org/wp-content/uploads/2023/09/ASEAN-JS-on-Climate-Change-to-the-UNFCCC-COP28.pdf

ASEAN. 2023b. "Chairman's Statement of the 13th ASEAN-United Nations Summit." Jakarta, Indonesia, September 7. https://asean.org/wp-content/uploads/2023/09/Chairmans-Statement-of-13th-ASEAN-UN-Summit_Final.pdf

ASEAN. 2023c. "ASEAN Leaders' Declaration on Sustainable Resilience." Jakarta, Indonesia, September 5. https://asean.org/wp-content/uploads/2023/09/ASEAN-Leaders-Declaration-on-Sustainable-Resilience.pdf

Belmont, P., and C. De Coning. 2022. "Coping with Complexity: Toward Epistemological Pluralism in Climate – Conflict Scholarship." *International Studies Review* viac055:1–29.

Bremberg, N., M. Mobjörk, and F. Krampe. 2022. "Global Responses to Climate Security: Discourses, Institutions and Actions." *Journal of Peacebuilding & Development* 17 (3):341–356. https://doi.org/10.1177/15423166221128180.

Busby, J., T. G. Smith, N. Krishnan, C. Wight, and S. Vallejo-Gutierrez. 2018. "In Harm's Way: Climate Security Vulnerability in Asia." *World Development* 112: 88–118. https://doi.org/10.1016/j.worlddev.2018.07.007.

Busby, J. W. 2021. "Beyond Internal Conflict: The Emergent Practice of Climate Security." *Journal of Peace Research* 58 (1):186–194. https://doi.org/10.1177/0022343320971019.

Caballero-Anthony, M. 2022. "The ASEAN Way and the Changing Security Environment: Navigating Challenges to Informality and Centrality." *International Politics*. https://doi.org/10.1057/s41311-022-00400-0.

Caballero-Anthony, M., and R. Emmers. 2006. "Understanding the Dynamics of Securitizing Non-Traditional Security." In *Non-Traditional Security in Asia: Dilemmas in Securitization*, edited by M. Caballero-Anthony, R. Emmers and A. Acharya, 1–12. Aldershot, UK: Ashgate.

Caballero-Anthony, M. et al. 2023. *Climate Change and Its Impact on Peace and Security in Southeast Asia*. Singapore: Centre for Non-Traditional Security Studies, S. Rajaratnam School of International Studies, Nanyang Technological University, 1–104. https://www.rsis.edu.sg/wp-content/uploads/2024/01/UN-Report-RSIS-Climate-Change-and-Its-Impact-on-Peace-and-Security-in-Southeast-Asia-Online-Version.pdf.

Chong, A. 2023. "Climate Change as Spatial Change: Impetus to Rethink State Obligations and Embrace Supra-Sovereign Knowledge." *Geopolitics* :1–26. https://doi.org/10.1080/14650045.2023.2225055.

Conca, K. 2019. "Is There a Role for the UN Security Council on Climate Change?" *Environment: Science and Policy for Sustainable Development* 61 (1):4–15. https://doi.org/10.1080/00139157.2019.1540811.

Cook, A. D. B., and S. Nanthini. 2023. "Mitigate, Adapt, Prepare: A Climate Security Agenda for the ASEAN Defence Ministers' Meeting." RSIS Policy Report, June. https://www.rsis.edu.sg/rsis-publication/nts/mitigate-adapt-prepare-a-climate-security-agenda-for-the-asean-defence-ministers-meeting/

Department of Defense (DOD). 2021a. *Climate Risk Analysis*, 1–18. Washington, DC: Department of Defense.

DOD. 2021b. *Climate Adaptation Plan*, 1–28. Washington, DC: Department of Defense.

DOD. 2022. *Climate Adaptation Plan 2022 Progress Report*, 1–10. Washington, DC: Department of Defense.

Elliott, L. 2012. "Human Security, Climate Change and Migration in Southeast Asia." In *Climate Change, Migration and Human Security in Southeast Asia. RSIS Monograph No. 24*, 1–12, edited by L. Elliott. Singapore: S. Rajaratnam School of International Studies.

ESCAP. 2022. "Asia-Pacific Riskscape @ 1.5 °C: Subregional Pathways for Adaptation and Resilience. Asia-Pacific Disaster Report 2022 for ESCAP Subregions, Summary for Policymakers." Bangkok: ESCAP. https://www.unescap.org/sites/default/d8files/knowledge-products/Asia%20Pacific%20Disaster%20Report%202022%20for%20ESCAP%20Subregions%20Summary%20for%20Policymakers_Rev.%206-4.pdf

Floyd, R. 2015. "Global Climate Security Governance: A Case of Institutional and Ideational Fragmentation." *Conflict, Security & Development* 15 (2):119–146. https://doi.org/10.1080/14678802.2015.1034452.

Foot, R., and E. Goh. 2019. "The International Relations of East Asia: A New Research Prospectus." *International Studies Review* 21 (3):398–423. https://doi.org/10.1093/isr/viy015.

Goodman, S, et al. 2021. "Climate Change and Security in the Arctic." The Center for Climate and Security (CCS), an institute of the Council on Strategic Risks (CSR), and The Norwegian Institute of International Affairs (NUPI). https://climateandsecurity.org/wp-content/uploads/2021/01/Climate-Change-and-Security-in-the-Arctic_CCS_NUPI_January-2021-1.pdf

Hameiri, S, and L. Jones. 2013. "The Politics and Governance of Non-Traditional Security." *International Studies Quarterly* 57 (3):462–473. https://doi.org/10.1111/isqu.12014.

Hauger, J. S. 2012. "Climate Change and Environmental Security in the Asia-Pacific Region: A Role for APEC?" In *Climate Change and Environmental Security*. Honolulu: DKI Asia-Pacific Center for Security Studies. https://dkiapcss.edu/wp-content/uploads/2012/09/Chapter2.pdf.

Jayaram, D., and M. C. Brisbois. 2021. "Aiding or Undermining? The Military as an Emergent Actor in Global Climate Governance." *Earth System Governance* 9:100107. https://doi.org/10.1016/j.esg.2021.100107.

Karki, R., J. Yu, S. Fetzek, M.-G. Manea, K. Ono, and N. Regaud. 2023. "Climate Security and Security Sector Governance in East Asia." In *Thematic SSG Brief*, 1–18. Geneva: DCAF.

Krampe, F., and M. Mobjörk. 2018. "Responding to Climate-Related Security Risks: Reviewing Regional Organisations in Asia and Africa." *Current Climate Change Reports* 4 (4):330–337. https://doi.org/10.1007/s40641-018-0118-x.

Krampe, F., R. Scassa, and G. Mitrotta. 2018. "Responses to Climate-Related Security Risks: Reviewing Regional Organizations in Asia and Africa." SIPRI Insights in Peace and Security, August.

McDonald, M. 2013. "Discourses of Climate Security." *Political Geography* 33:42–51. https://doi.org/10.1016/j.polgeo.2013.01.002.

McDonald, M. 2018. "Climate Change and Security: Towards Ecological Security?" *International Theory* 10 (2):153–180. https://doi.org/10.1017/S1752971918000039.

McDonald, M. 2021. *Ecological Security*. Cambridge: Cambridge University Press.

MINDEF. 2021. "Singapore Affirms Commitment to Multilateral Cooperation and an Open and Inclusive Regional Security Environment." News Release, November 10.

NADI. 2019. Chairman's Report of the Track II Network of ASEAN Defence and Security Institutions (NADI) Workshop on "ASEAN Defence Cooperation in Dealing with Climate Change." https://www.

rsis.edu.sg/nadi/pdfs/climatechange_1/Chairman's%20Report%20of%20the%20NADI%20 Workshop.pdf. Retrieved via https://www.rsis.edu.sg/nadi/nadi_climatechange_1.html

NATO. 2021. "Climate Change and Security Action Plan." June 14. https://www.nato.int/cps/en/natohq/official_texts_185174.htm

New Zealand Ministry of Defence. 2018. *The Climate Crisis: Defence Readiness and Responsibilities*. Auckland: New Zealand.

Prayitno, H. 2021. "AUKUS and the Role of ASEAN Centrality in Managing Regional Security Affairs." *Technium Social Sciences Journal* 26:678–688.

Prizzia, R., and J. Levy. 2018. "Towards Climate Security and Sustainable Security in the Asia-Pacific Region." In *Asia-Pacific Security Challenges: Managing Black Swans and Persistent Threats*, edited by A. J. Masys and L. S. F. Lin, 41–64. Cham: Springer International Publishing.

Remmits, F., and L. Birkman. 2020. "The Face of Climate Insecurity: A Surge of Tropical Storms Strike Megacities in Asia and the Pacific." HCSS Security, November. https://hcss.nl/wp-content/uploads/2020/11/Megacities_Snapshot.pdf

Seah, S., and M. Martinus. 2021. *Gaps and Opportunities in ASEAN's Climate Governance. Trends in Southeast Asia*, 1–29. Singapore: ISEAS Yusof-Ishak Institute.

Shidore, S., S. Fetzek, and R. Fleishman. 2021. "The Link Between National Security and Climate Change in Asia." The Cipher Brief, April 21. https://www.thecipherbrief.com/article/asia/the-link-between-national-security-and-climate-change-in-asia

Starr, J. R. 1991. "Water Wars." *Foreign Policy* 82 (Spring):17–36.

Tangney, P., C. Nettle, B. Clarke, J. Newman, and C. Star. 2021. "Climate Security in the Indo-Pacific: A Systematic Review of Governance Challenges for Enhancing Regional Climate Resilience." *Climatic Change* 167 (3-4):40. https://doi.org/10.1007/s10584-021-03197-8.

Thomas, K. A., and B. P. Warner. 2019. "Weaponising Vulnerability to Climate Change." *Global Environmental Change* 57:101928. https://doi.org/10.1016/j.gloenvcha.2019.101928.

Thomas, M. 2022. "Climate Change and Military Forces." In *The Geopolitics of Climate and Security in the Indo-Pacific*, edited by R. Glasser, C. Johnstone, and A. Kapetas. ACT, Australia: APSI and KAS.

UN. 2021. *Asia and the Pacific SDG Progress Report 2021*. Bangkok: United Nations. https://www.unescap.org/sites/default/d8files/knowledge-products/ESCAP_Asia_and_the_Pacific_SDG_Progress_Report_2021.pdf

Cascading risks in a social-ecological system: the South China Sea disputes

Maria Ortuoste

ABSTRACT
The situation in the South China Sea (SCS) is the epitome of a wicked problem where issues cross state boundaries and domains, and where stakeholders range from fishers to governments to shipping industries. This article frames the SCS as a social-ecological system that is composed of biological and social subsystems. It explains how, for over 70 years, global political and economic structures have conditioned state responses – territorialisation, militarisation, extraction, and exploitation – that, in turn, have led to cascading risks for further environmental degradation and conflict at different levels of community. This article uses the grammar of riskification, rather than securitisation, because the crisis in the SCS requires governance rather than military action. Any code of conduct needs to emphasise stewardship rather than ownership and to be climate-centred rather than threat-centred, in order to address the very real risk of the collapsing biological subsystem upon which the social subsystem ultimately depends.

Introduction

It has become commonplace for policymakers to say that climate change will affect international security; yet connecting the dots from climate science to political action remains elusive. An Institute of Southeast Asian Studies or ISEAS (2021) survey of Asian policymakers and analysts showed that while 54% of the respondents considered climate change a 'serious and immediate threat to the well-being of their countries', only 20.8% of respondents expressed concern for depleted fish stock and biodiversity loss in the South China Sea (SCS) which is due to coral bleaching, a byproduct of increased sea temperatures due to climate change (Xiao et al. 2022). This depletion leads to more incidences of illegal, unreported and unregulated fishing (IUUF) which is acknowledged as a regional security problem.[1] This disconnect is due not only to the narrow interpretation of national interests but also to *how* those interests are shaped by existing political and economic structures.

This current situation in the SCS is a wicked problem which crosses levels of analysis, economic and social sectors, borders and jurisdictions, and even disciplines. The co-constitutiveness of nature and human society is modelled as a social-ecological system composed of biological and social subsystems. In the SCS, the accumulation of human activity has led to cascading risks for environmental degradation and climate change, but there are avenues for cooperation. A holistic and transdisciplinary perspective is necessary at a time when '[s]ecurity is now about navigating contesting ideas of how affected, insecure communities can be made more secure' (Caballero-Anthony 2017, 125). The grammar of 'risk' highlights the role of governance, rather than the role of the military, in addressing the root causes of environmental damage and climate change.

This article begins with a short review of climate security and conflict pathway approaches which come close to, but never fully explain, the co-constitutive relationship between nature and society. It then discusses the biological and social subsystems that make up the social-ecological system of the SCS. For over 70 years, global political and economic structures have conditioned state responses. The territorialisation of the oceans, the militarisation of outposts and the use of resource nationalism increase tensions, and preparations for conflict lead to more environmental damage. Existing Association of Southeast Asian Nations' (ASEAN) mechanisms, while limited, can become foundations for a climate-centred code of conduct in the South China Sea.

Wicked problems

Climate security and conflict pathways

Policymakers, scholars, think tanks, people's organisations, and communities of practice argue that environmental degradation and climate change can drive or aggravate conflict at local, national, and international levels, especially if there are pre-existing systemic inequalities and violence (McDonald 2013; Busby 2021).[2] Several studies discuss how the interaction of climate change, conflict drivers (economic, social and political), and institutions can lead to risks for violence or the outbreak of conflict at interpersonal, intrastate and interstate levels (Burke et al. 2015; Blecher 2022; Kalkavan 2017). Other studies have identified the direct and indirect pathways by which climate-induced extreme weather patterns can lead to environmental crime, riots and conflicts (Ruttinger et al. 2022; Sekiyama 2022; Theisen 2017; Corry 2012). Fleishman (2022) identified four possible scenarios for regional conflict or cooperation in the SCS based on countries' levels of political fragmentation and fluctuating food prices, recognising that 'key socio-political, technological, demographic, diplomatic, military and economic drivers' will shape the 'climate security risk landscape in the region'.

These approaches, however, are still problematic for at least two reasons. The need to secure the state over people is still a dominant tendency. For example, the International Military Council on Climate and Security (IMCCS) defines climate security as the point when

> ... human security risks spill over into higher-order security risks, such as political instability, conflict, major natural disasters involving significant military and humanitarian responses, mass displacements of peoples, and threats to critical resources and infrastructure (2021, 19).

Yet they neglect to mention that these security risks arose from the very acts taken to ensure state and economic security. As Buzan, Wæver and de Wilde (1998, 76) observed, it is difficult to escape from the paradox whereby the '... only way to secure societies from environmental threats is to change them'. Even when there is an effort to identify how a range of actions – prevention, emergency responses, and peace building – fall between climate change on one end, and (hard) security interventions on the other end, it still involves a 'politically conservative bias' that leaves in place the very structures which created the challenges in the first place (Wæver 1995, 64). It is, therefore, unsurprising that by the end of 2021, the world's defence budget had passed $2.3 trillion (SIPRI 2022), while climate finances fell short of the paltry $100 billion annual commitment (Green Climate Fund n.d.).

An us-versus-them approach accompanies this status quo bias. For example, the displacement of peoples due to climate change is often articulated as threats, justifying military responses which do not mitigate but rather exacerbate cross-border ecological problems (Deudney 1999, 193–94). While governments have already identified mitigation, adaptation, prevention and response actions, they have yet to make deeper societal, political and economic transformations.

Wicked problems

A wicked problems approach allows for a less siloed perspective than the abovementioned studies. The concept has been applied to the study of climate change and conflicts where multiple factors interact in non-linear ways, leading to an ever-evolving situation which makes 'solutions' ever-moving targets, difficult if not impossible to achieve (FitzGibbon and Mensah 2012; Garba 2022). Sam Bateman stated that maritime security in the Asia-Pacific is a wicked problem because of fundamental differences between stakeholders who are each convinced of the 'correctness of their positions', especially in the SCS (Bateman 2011). The situation in the SCS fits the description of wicked problems as 'complex and messy social problems which are difficult or impossible to solve because of the incomplete, contradictory, and constantly changing needs of multiple actors in various situations'. The issues are co-constitutive as they are embedded in 'a system of interrelated problems [where solutions] can seldom be obtained by independently solving each of the composed problems' (Hou, Li, and Song 2022, 303).

One advantage of the wicked problem approach is its openness to combining multiple perspectives. The diverse issues in the SCS require transdisciplinarity which 'revolves around real-world experiences, without regard for line-specific understandings … [and]involves multiple disciplines in order to connect new knowledge and deeper understanding to real life experiences' (Hou, Li, and Song 2022, 303). That is, it is best to view social and natural sciences as complementary simply because nature does not respect disciplinary or bureaucratic divides. In the case of climate change, it is necessary to look at the 'political, economic and social structures that serve to both separate people from the environment and give rise to processes of environmental change' (McDonald 2013, 48).

Purely geostrategic studies tend to assume that humans can fully control natural processes when, in fact, social systems are dependent on the biosphere for survival (see Emmerson 2020; Guan and Skogan 2007; McPhail et al. 2020). These studies also evince blindness to the environmental consequences of war, military preparedness, and resource extraction. In other words, the very methods used to guarantee state, and even human, security are in fact harming them in the long run. Conversely, purely scientific approaches can be so highly specialised that they can miss system-wide linkages (see Zhang et al. 2021; Yuan et al. 2019; Chen et al. 2023; and Liu et al. 2023). In addition, science itself can be manipulated for a state's own interests. One scholar, for example, observed that 'China has been reluctant to let outside scientists conduct fish stock assessments, leaving the international community partially blind to the scope of the problem' (Abraham et al. 2020).

A more important reason to use the wicked problems framing is ethical. Introduced by Rittel and Webber (1973), the wicked problem concept rejects rational-systems or data analysis approaches to policy development, 'arguing instead that all the professions concerned with social analysis and planning should focus on understanding the aspirations and values of the people rather than developing expert-led comprehensive plans' (Head 2022, 25–26). By clarifying the interests, roles, capabilities and vulnerabilities of various stakeholders, it would be possible to take more meaningful steps towards addressing problems. Moreover, it is a way to ensure that unequal structures that make certain populations, like small-scale fishers, vulnerable are being remedied.

Nevertheless, the wicked problem concept itself has been criticised for its ontological preoccupation with dichotomy, ie whether a problem is tame or wicked. This article does not subscribe to that division; rather, it argues that developments in the SCS are better seen as falling along a spectrum of intractability where risks and opportunities co-exist, and where one may be weightier than another at different points in time (Turnbull and Hoppe 2019; Head 2022).

Systems and cascading risks

Wicked problems cannot be addressed by 'approaches short on explicating the complex interconnections of the multiple causes, consequences, and cross-scale actors of the problem' (Fitzgibbon and Mensah 2012, 1). Systems frameworks help to identify interactions and dynamic feedback especially when the 'cumulative totality' of multiple issues 'are beginning to interact in all sorts of unpredictable synergies' (Dalby 2014, 4).

The SCS is a social-ecological system which consists of

> two interacting subsystems: the biological (epidemiological ecosystem) and the social (social and economic conditions of life of the society) subsystems where the biological subsystem plays the role of the governed object and the social acts as the internal regulator of these interactions.[3]

The phrase 'acts as' is of particular importance because the social subsystem can never fully control environmental consequences. The definition also captures a key feature of the Anthropocene era, namely the substantial alteration of the Earth due to the accumulation of human activities.

Biological subsystem

Arguably, there is only one biological system, just as there is one global ocean which has been conceptually divided into different ones. Nevertheless, scientists have been using the term large marine ecosystem (LME) which is defined 'as near coastal areas where primary productivity is generally higher than in open ocean areas' (Global Environment Facility, n.d.). LMEs transcend political jurisdictions and better convey the connectivity and flow of nature. The SCS is one of 66 LMEs in the world. It measures more than 3 million km^2 and is an extremely rich area.[4] Its total reef area is 12,000 km^2, which is almost 5% of the world's total reef surface area.[5] The area's reefs are part of the Coral Triangle where 76% of the world's coral species and 37% of the world's coral reef fish are found (ADB, n.d.). The South China Sea's inshore fishing area measures 890,000 km^2 and from 1950 to 2014, the annual average catch was 7.76 million tonnes, which was twice the annual average catch from the larger Arabian Sea LME (Sumaila 2019). It is also estimated to hold around 7 billion barrels of proven oil reserves and 900 trillion cubic feet of natural gas (Xu 2009).

Apart from these resources, there is heavy maritime traffic in the SCS, from fishing boats to ferries to cargo. The Malacca and Singapore Straits are historical trade routes upon which Indigenous seafaring communities depend. Around 271 million people depend on the LME for daily subsistence and livelihoods. This biological subsystem is currently in distress, as will be discussed in the next section.

Social subsystem

Large marine ecosystems (LMEs) challenge the very notion of sovereign statehood. Political actors in this social subsystem are organised according to norms of sovereignty which imagine a clear separation between domestic and international spheres, and where territorial space is supposed to be inviolable. Yet this is a territorial trap where the maritime space disappears, or is of secondary status, along with the 'dichotomization of land and maritime space' (Wirth 2018, 35; see also Agnew 1994). This limitation becomes inherent in the tasks of governing, namely territorialisation, militarisation, extraction, trade and maritime transport.

This means that governance practices on land are imposed on the maritime domain. One such practice is territorialisation. Six states claim all or part of the SCS waters and maritime features – Brunei, China, Malaysia, Philippines, Taiwan and Viet Nam. Indonesia claims the Natuna Islands which is at the south-eastern end of the SCS, and not the Paracels or Spratlys islands. They benefit from resource extraction, trade and shipping-related industries. Simultaneously, there are geopolitical relationships involved – Thailand, the Philippines and Taiwan are long-time allies of the US, which is in a power competition with China. At the same time, these countries are engaged in multilateral associations such as ASEAN, the ASEAN Regional Forum (ARF), and the Asia Pacific Economic Cooperation (APEC) forum.

States are not the only stakeholders in the SCS. Stakeholders include seafaring and coastal communities, national governments and their military and civilian agencies, private actors (fishing, shipping, tourism and petroleum industries), multinational corporations, and extra-regional actors such as allies and international organisations. The capabilities of the stakeholders vary – at the lowest end are the fishing community who are the poorest in their countries and directly affected by extreme weather conditions (IMCCS Expert Group 2021, 21). While governments do pass laws and enforce rules, it should not be presumed that governments are the ultimate decision makers; private industries and multinational corporations can, at times, wield economic and financial influence sufficient to sway national policies in their favour.

Embedded systems and cascading risks

As seen in Table 1, these stakeholders' actions – namely territorialisation, militarisation, extraction, movement and trade – vary in scale, and in the scope and severity of their

Table 1. Stakeholders' interests in and environmental impact on the South China Sea.

Stakeholders	Interests	Negative environmental effect due to
Seafaring communities, coastal communities	Food/daily subsistence Jobs in various industries Survival/home	Destructive fishing practices
Fishing industry	Profits and access Less regulation	Fishing equipment and destructive fishing practices Harvesting of endangered species Frequency and efficiency of fishing
Oil industry	Profits and access Less regulation	Underwater sound and light disturbing marine habitats Increased vessel traffic Oil spills
Shipping and multinational corporations	Profits and access Less regulation	Emissions Run-off from land-based factories
Tourism industry	Profit, access to beaches	Trash Sewage
Countries' governments	Territorial integrity Economic development Governance and control Protection	All of the above Emissions from naval and coast guard patrols and operations Construction of outposts
Allies	Access Sphere of influence	Military bases Emissions Pollution (noise and waste) during military exercises

Note: This table only includes the stakeholders mentioned in this article.

maritime impacts. They also have different capacities to affect policy, and varying exposure to climate shocks. Thus, there can be no one definitive formulation of the problem with each stakeholder having their own preferred explanation and/or solution to the problem. Because these social and ecological issues are worsening over time, potential solutions need to be cognisant of how one 'solution' begets another problem.

Figure 1 shows the SCS social-ecological system as embedded within wider social-ecological systems. The arrows inside the innermost circle represent the dynamic interaction and co-constitution of the two subsystems. Arrows across systems similarly represent this mutual constitutiveness and 'cascading impacts and implications across linked human-environment systems', or '"cascade[s]" across and between multiple domains' (Cradock-Henry et al. 2020).

The interrelated issues in the SCS are geopolitics and the global economy, environmental degradation and natural disasters, competitive and cooperative interstate relations, and various domestic crises within littoral countries. These issues are unfolding simultaneously at different levels through various processes – territorialisation of the maritime area, militarisation of outposts, resource extraction, and increased maritime traffic. The unfolding of these processes leads to endless 'causal chains' where actions in one area will have repercussions on other issues. Over time, the situation becomes more complicated as piecemeal actions are, at best, only good in the short term and, at worst, generate unintended negative consequences or quickly become irrelevant because of rapid changes in stakeholders' needs.

Nevertheless, this article consciously avoids the term *threat* and follows a grammar of 'riskification'. Securitisation constructs scenarios of direct harm to a valued referent object and, therefore, legitimises exceptional measures for survival. Riskification, in contrast, discusses 'constitutive causes [that] frame entire situations' and legitimises 'precautionary

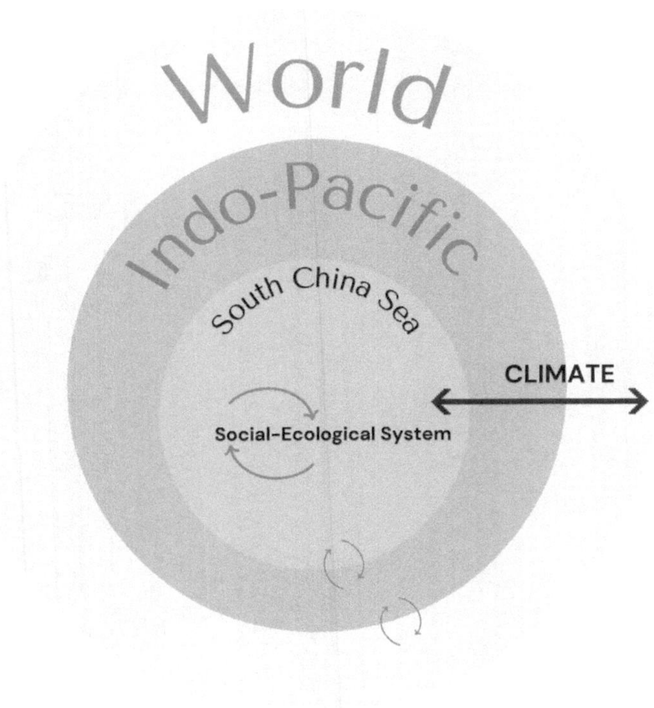

Figure 1. Embedded social-ecological systems and cascading risk.

measures' or 'a plan of action to *govern* the conditions of possibility for harm' rather than defaulting to military solutions (Corry 2012, 246–48).

The South China Sea as a social-ecological system

The case study is divided into three periods marked by distinctive trends in geopolitics, economics and stakeholder actions in the SCS. Figure 2 shows that as resource extraction, maritime transport, territorialisation and militarisation intensify over time, there is a corresponding increase in global CO_2 emissions, degradation of marine habitats, and depletion of fish stocks. Regional forums have also developed over time, and negotiations for a code of conduct on the SCS are continuing but they have had marginal impact on ameliorating environmental conditions or resolving disputes in the area.

World War II to 1989: political and economic transitions

The key geopolitical trends during this period were the independence of colonies and the Cold War, while Globalisation 2.0, or 'the great acceleration' of trade, was occurring.[6] China and Southeast Asian states were trying to secure their land territories and maritime zones in order to extract and exploit resources, as well as to assert their sovereign statehood. The negotiations for the United Nations Convention on the Law of the Sea (UNCLOS) set the stage for expansive claims over exclusive economic zones (EEZs). By the mid-1980s, five claimants had occupied 26 features in the SCS.[7]

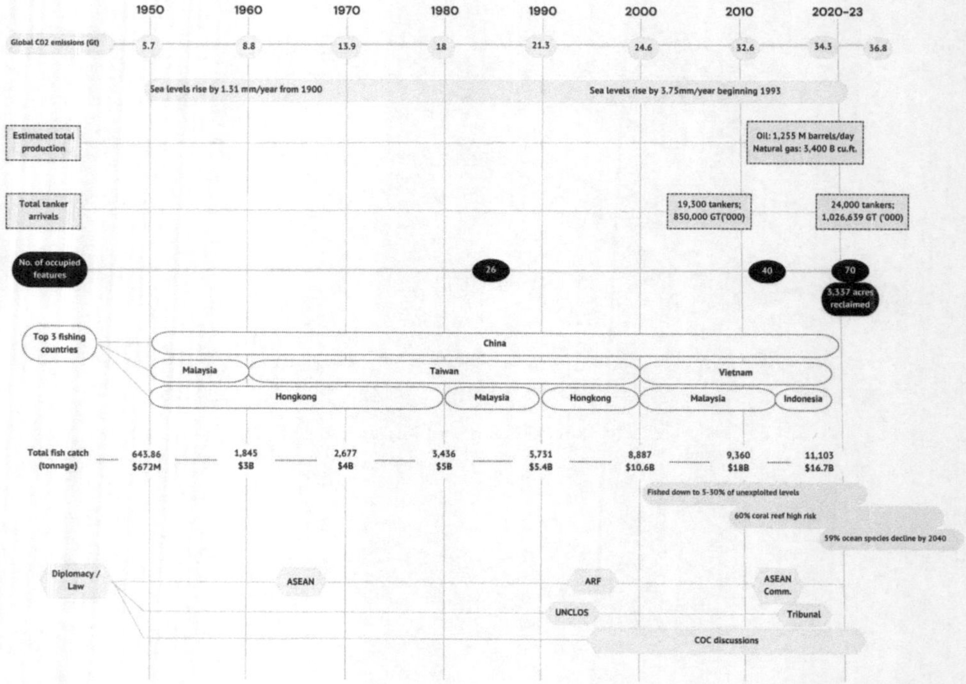

Figure 2. Social-ecological system evolution, 1950–2023. ASEAN Comm: ASEAN Community; COC: code of conduct; ASEAN: Association of Southeast Asian Nations; ARF: ASEAN Regional Forum; UNCLOS: UN Convention on the Law of the Sea; GT: gross tonnage.

Sources: Data from Pauly, Zeller, and Palomares (2020) for total fish catch and top three fishing countries; US Energy Information Agency (2024) for estimated 2011 production of oil and natural gas; Singapore Maritime and Port Authority (2023) for total vessel arrivals in Singapore's ports; AMTI (2022) for occupied features; NDTV World News (2022) for sea level rise; Bush and Marschke (2016) for 'fished down' data; Burke et al. (2011) for coral reef decline; and International Energy Agency (2023) for global CO_2 emissions.

Littoral countries were positioned to take advantage of this trend. By the mid-1980s, Singapore, Malaysia, Indonesia and Thailand had opened their modernised ports to container ships and tankers. This helped increase 'developing Asia's' trade in goods and services from only 25% of the regional gross domestic product (GDP) in 1970 to over 60% by 1989 (IMF Staff 2008). Oil and natural gas fields, many near the coasts, boosted the economic growth of Brunei, Indonesia and Malaysia.[8] Fish catch in the SCS increased dramatically from only 643.86 tonnes in 1950 to 5731 tonnes by 1990.

China, the Philippines and Viet Nam engaged in direct island occupation in search of energy and fishery resources, according to the Maritime Awareness Project.[9] These structures were not fully militarised, but violence did occur twice between China and Viet Nam, in the 1970s and late 1980s (Chubb 2022, 12). These skirmishes did not hinder the movement of the US Navy, which was able to 'freely and routinely' operate 'throughout the Indo-Pacific in support of treaty and partnership commitments and American interests' (Blair 2016). The USSR was not a competing naval power in the area, regional countries were preoccupied with internal security issues, and US allies supported this presence.

These economic activities, the oil extraction processes, and the construction of outposts, in addition to maritime traffic and land-based industrial activities contributed to the pollution and degradation of marine and terrestrial environments. They also contributed to the increase in global CO_2 emissions from 5.7 billion tonnes in 1950 to 21 billion tonnes in 1990. Sea levels in the SCS rose by 1.31 mm per year (NDTV World News 2022). Internationally, many scientists were already sounding the alarm about the warming effect of greenhouse gases, but most regional countries were more concerned about economic growth than conservation.[10]

1990 to 2015: cascading risks, few opportunities

During this period, the biological subsystem was already under stress. By 2015, global CO_2 emissions had reached 33 billion tonnes, which roughly correlates with the increasing share of exports in the world's GDP (Vanham 2019).[11] Developing Asia's trade in goods and services grew exponentially to become 95% of regional GDP, while intra-ASEAN trade increased 'from 29% of total ASEAN trade in 1980 to 41% in 2009' (IMF staff 2008; Xu 2009). That same year, the US National Oceanic and Atmospheric Association (NOAA) reported that 'commercial vessels – freighters, tankers and cruise ships – generate[d] enough air pollution to pose "a significant health concern for coastal communities"', especially in East Asia (McClatchy Newspapers 2009). This is not surprising because '[m]ore than half of the world's top ten shipping ports are … located in and around the SCS, according to the International Association of Ports and Harbors' (Xu 2009). Around one-third of world maritime oil trade and one-fifth of world maritime oil supply passed through the Malacca Straits in 2013 (IISS 2015, 34).

Pollution from maritime traffic is compounded by resource extraction. With coastal oil fields maturing, littoral countries doubled their efforts to find oil and natural gas reserves in the SCS.[12] More problematic was the accelerated degradation of coral reefs and depletion of fish stock in the area due to overfishing. Asia became the main producer of aquatic products by 1993, and the Food and Agriculture Organization of the United Nations (FAO) estimates that by 2015, the volume of fish catch in the SCS was roughly twice more than the volume in 1990, and 12% of global catch (FAO 2022). China, Viet Nam, Indonesia, Thailand and the Philippines were among the world's top 10 aquaculture producers (FAO 2020). In addition, more than half of the world's fishing fleet is positioned in the SCS (Sumaila and Cheung 2016, 8). But as early as 2008, fish stock in the SCS was reportedly collapsing, and about 60% of the coral reefs were considered to be at high risk.[13]

The immediate effects of this environmental damage were felt by coastal communities, which are the most vulnerable to reef loss and worsening weather conditions.[14] Small-scale fishers reported lower catches per unit effort, accompanied by a 'significant decline' in the 'size and diversity of fish species' between 1998 and 2008, and by 2013, small-scale catch had declined to only 35% of total fish catch (Bush and Marschke 2016; Teh and Pauly 2018). Researchers have found that '[s]ocial unrest between fishers is real and is enhanced by general fishing declines, aquatic and coastal exploration, exploitation, and expansion' (Bush and Marschke 2016, 231). Indigenous seafaring communities (such as the Orang Laut, Sama Bajau, and Bugis) lost not only their livelihood but also a valuable part of their culture and identity (Gaynor 2005).

These factors partially explain, and are used by governments to justify, hyper-territorialisation and militarisation in the SCS. Littoral countries submitted the limits of their EEZs and continental shelves to the UN, and have been filing diplomatic protests against each other ever since UNCLOS came into effect in 1994. By 2015, claimants had occupied 40 features and constructed fortified structures to accommodate bigger ships and aircraft, thereby ensuring sustained military presence (McManus et al. 2016). The Maritime Awareness Project found that patterns of coercive measures changed in this period. From the 1990s to 2002, the Philippines took coercive measures against Chinese trawler crews and maritime militia; from 2000 to 2010, tensions were mainly between Viet Nam and China; and beginning in 2010, Chinese coercive actions were against the Philippines (Chubb 2022). Alarming episodes included China's occupation of Mischief Reef in 1995, Viet Nam's accusations that Chinese fishing boats cut cables from an oil exploration vessel operating within its own EEZ (2011), and the Scarborough Shoal stand-off between the Philippines and China (2012).[15]

Militarisation was also a result of concerns about US military drawdown at the end of the Cold War and its focus on the war on terror. Having experienced significant economic growth, Northeast and Southeast Asian countries implemented defence modernisation programmes that were now orientated outward into the maritime sphere (Ortuoste 2018).

Chinese military expenditures were also growing by leaps and bounds, and by the 2010s, China had launched its first aircraft carrier, declared an air defence identification zone (ADIZ), and embarked on aggressive land reclamation in their occupied features to accommodate more ships – fishing boats, and ships under the command of their coast guard, the maritime militia, and their navy. These actions would not go unchallenged. The Obama administration began to articulate its 'strategic rebalancing' to the Indo-Pacific, and by 2015 it was conducting Freedom of Navigation operations (FONOPs) with US ships sailing near Chinese-built structures.

Within countries, there was increased use of resource nationalism due to the growing frequency of IUUF.[16] The governments of China, Indonesia, Viet Nam and the Philippines encouraged their fishers to relocate to islands (such as Pag-Asa or the Natunas) and to fish near, or in, disputed waters. Unfortunately, the fishers do not always get enough financial support, and they are subject to arrest and incarceration by neighbouring countries, even if this is not allowed under UNCLOS. Relocating people can also cause internal tensions, such as in Indonesia where the local population in the Natuna Regency disapproved of the fishing practices of these 'migrants' (Oceans Inc 2021). Resource nationalism is also reflected in popular protests and demonstrations. For the most part, these protests are loud but non-violent; but in 2014, 21 people were killed and around 100 injured when Vietnamese 'set fire to industrial parks and factories, [and] hunted down Chinese workers' as a reaction to China's stationing of an oil rig in disputed waters (Hodal and Kaiman 2014).

During this same period, the establishment of ARF, its working groups, and the so-called ASEAN Plus meetings has allowed for greater conversations and proposals for confidence-building, preventive diplomacy, and cooperation both in the SCS and on environmental issues. One of the first forums was the Indonesian-sponsored track-two forum called Managing Potential Conflicts in the South China Sea (MPC-SCS) which began in the 1990s and continues today. Its uniqueness stems from the inclusion of Taiwan, and from its focus on functional cooperation rather than on sovereignty issues.

ASEAN slowly began articulating some norms on environmental protection and sustainability. In their Hanoi Plan of Action (1998), ASEAN hoped to create a regional action plan to protect the marine environment, as well as to improve coordination for integration protection and management of coastal zones. This plan, however, took a backseat because of the haze problem. However by 2007, the ASEAN Political-Security Community Blueprint identified 'strengthening cooperation on disaster management and emergency response' in the face of climate change, and the organisation created the relevant ministerial bodies (Caballero-Anthony 2017, 134). The organisation also created networks on environmental governance such as the ASEAN Social Forestry Network (2005), the ASEAN Regional Knowledge Network on Forest and Climate Change (2008), and the ASEAN Climate Resilience Network (2013) (Salazar 2023).

ASEAN, however, hesitated to address the issue of China's increasingly aggressive actions in its formal agenda in the 1990s because of their growing economic links with China, the perceived need to retain US maritime presence, and ASEAN's lack of dispute resolution mechanisms. The 1992 Declaration on the SCS called for peaceful resolution of disputes, the exercise of restraint, and cooperation on areas such as marine scientific exploration and environmental protection. Yet given the unabated use of the oceans, this suggestion rang hollow. In reaction to the 1995 Mischief Reef incident, that year's Joint Communique of the ASEAN Ministerial Meeting called for self-restraint among the claimants, and to use the Indonesia-led workshops to establish trust and identify areas for cooperation. After seven years, in 2002, ASEAN countries and China signed the Declaration on the Conduct of Parties in the SCS, laying out the principles on which an actual code of conduct would be based. The significance of this document was marred by the length of time it took to simply identify those principles. In the interim seven years, claimants occupied more features, built up their outposts, and modernised their militaries. Ongoing negotiations did not prevent scientific exploration, selling concessions to oil blocks, the arrest of fishermen and, beginning in the 2010s, China's increased harassment of fishing boats and coast guard vessels of other claimants.[17]

China's actions prompted several countries to improve their strategic partnerships not only with the US but also with Japan, South Korea, and Australia, who were beneficiaries of open trade routes. Unfortunately, China could seize upon these defensive actions as proof of malign intentions and use its economic power as leverage against other claimants. Without any regional dispute resolution mechanism, or a unified ASEAN statement, or even concrete security guarantees from its main allies, the Philippines instituted arbitration proceedings against China in 2013. It sought opinions regarding their maritime entitlements and China's actions in disputed waters. Coincidentally or not, China agreed to restart discussion on a code of conduct in that same year, with the parties agreeing to a 'framework' by May 2017 (Xinhua 2021).

2016–2022: temperatures rise

Like the rest of the world, the Indo-Pacific is experiencing extreme heat, deadly storms and floods, an unabated increase in global emissions, and unprecedented species extinction. The SCS is now considered to be 'fully- or over-fished', but as seen in Figure 2, fish catch continues to increase (Baker 2016). IUUF also keeps growing, with China, Indonesia, Taiwan and Viet Nam considered among the top 20 countries most at risk for IUUF.[18]

By 2020, global CO_2 emissions had reached 36 billion tonnes, and maritime traffic in the SCS certainly contributed to these emissions because an estimated '50% of global oil tanker shipments pass through the SCS, which sees three times more tanker traffic than the Suez Canal and more than five times that of the Panama Canal' (Teh and Pauly 2018). The resulting ocean warming and acidification, as well as oil exploration and production, overfishing, and land reclamation activities, destroy coral reefs (Basu and Chaturvedi 2021; Solomon 2016).

By 2022, six claimants had installed 90 outposts on 70 features in the SCS. The total reclaimed area reached 3336.56 acres, 95% of which belonged to by China (AMTI 2022). The environmental effects of this reclamation will be felt over at least 1200 km^2 of the Pacific Ocean, where 'commercially vital fish stocks will lose their habitats and the sedimentation released by dredging will degrade the ocean's ability to detoxify waste that can impact climate' (Smith et al. 2019). Reefs were already destroyed by giant clam extraction, but 'filling operations [make] reef recovery impossible in such areas' (McManus et al. 2016).

Militarisation likewise contributes to environmental degradation and tense relations. Military outposts generate waste, and the increasing frequency of resupply and patrol missions is also detrimental to marine life. This presence facilitates China's coercive actions, often labelled 'grey zone tactics', such as increased trawler presence in the area, and the deployment of maritime militia, coast guard and naval vessels. A 2019 study that used advanced imaging technology showed that in 2018, around '300 ships [were] anchored at two [Chinese-occupied] reefs at any given time' (Poling 2019).[19]

The Asia Maritime Transparency Initaitive or AMTI (2020) reports that China has effectively extended its force projection by some 900 miles with radar, anti-ship and anti-air multiplatforms, and combat aircraft deployed to Woody Island, Fiery Cross Reef, Mischief Reef and Subi Reef. To counter China's pursuit of 'a sphere of influence in the Indo-Pacific', the US released its Indo-Pacific Strategy (White House 2022). One of its objectives is to maintain a 'free and open Indo-Pacific' by reinforcing deterrence. The US has seemingly abandoned ambiguity, declaring that its treaty commitments are 'ironclad', and conducting FONOPs in the area on a quarterly basis, inviting more participants to its multilateral naval exercises, and negotiating access to more of its allies' bases (Heydarian 2023; Mahadzir 2022; Vergun 2023).

Such actions, however, increase risks for both social and ecological subsystems. The Rim of the Pacific (RIMPAC) exercises, for example, help improve the capacities of the participating countries' forces and are a show of deterrent force; but the sonar and explosive devices used in the exercises harm marine life and also add to carbon emissions, according to environmental groups (Solomon 2016).

Small-scale fishers continue to be disadvantaged by these actions. Their share of total fish catch in the SCS dwindled from 85% in the 1950s to only 25% (Fabinyi et al. 2022). In addition, several fishers report suffering trauma from harassment by China's maritime militia (Oceans Inc 2021; Abraham et al. 2020). These direct encounters heighten resource nationalism. One scholar observed that China's use of force 'creates resentment towards the country among Vietnamese fishermen' which might encourage some people to join maritime militias or similar entities that are organised by their own governments (Abraham et al. 2020).[20] Relocating small-scale fishers or hiring them as militia does not, however, solve persistent challenges such as 'market access; financial services; livelihoods; poor access to public services such as health care, clean water, and sanitation; social protection; political and economic marginalization; gender inequity; [and] natural disasters' (Ferrer et al. 2021, 100).

Notwithstanding these challenges, ASEAN and China did agree on a Framework for a code of conduct in 2017, and a Single Draft Negotiating Text (SDNT) in 2018. Some of the promising provisions include preventing and managing disputes, monitoring the code's implementation, and the adoption of the 'duty to cooperate' clause from UNCLOS. The Philippines and China even submitted a joint proposal to provide 'just and humane treatment of all persons who are either in danger or in distress' (Thayer 2018).

Remaining challenges, however, are significant: what is the geographic scope of the code; will the code be legally binding; will there be a dispute settlement mechanism? Continued militarisation and resource nationalism complicate negotiations. With China having the upper hand in terms of occupation, it can afford to delay proceedings as it had done in 2002.

Due to China's expansive claims over, and its aggressive actions in, the SCS. The Philippine government filed an arbitration case against China under Annex VII of UNCLOS in 2013. In 2016, the International Tribunal of the Law of the Sea (ITLOS) ruled overwhelmingly in favor of the Philippines. But China's declaration that this 2016 Award is 'null and void' precludes mentioning that award in a code. This is unfortunate because the Award clarified several UNCLOS provisions (such as what constitutes military action, the role of Chapter XII, the regime of islands, and traditional fishing areas), which could define parameters for cooperation.

But the Award can still be useful. One of the Framework's objectives is to establish 'rules-based framework' for the code. The treaties mentioned in the Award could be these rules. ASEAN members and China are parties to the International Regulations for Preventing Collisions at Sea (COLREGS) and the Convention on International Trade in Endangered Species of Wild Fauna and Flora (CITES), which came into force *before* the Award. Moreover, in 2016, ASEAN and China agreed to apply the Code for Unplanned Encounters at Sea (CUES) to the South China Sea (Storey 2017). Since many provisions in CUES are based on COLREGS, the latter's provisions can be included in the code. The negotiating parties could use the code of conduct to reiterate their commitment to those treaties, or to create new, or reconfigure existing, mechanisms to address environmental concerns. CUES and COLREGS, for example, encourage the establishment of hotlines among signatories; perhaps parties to the code could establish 'climate hotlines' to support disaster response.

Yet a caveat needs to be made. The negotiating parties cannot just pick up where they left off, because circumstances are very different from 2002. Territorialisation and militarisation will probably not be stopped; constructions will most probably not be removed; and resource nationalism could prevent some compromises. But there is an urgent need to deal with the issues that contribute to climate change. Thus, any code of conduct or similar instrument has to be climate-centred.

The parties could certainly identify steps towards demilitarisation, but before then, they need to implement various proposals for practical environmental cooperation raised in various working groups of the ARF and ASEAN. ASEAN members have successfully cooperated with non-members in other multilateral endeavours 'such as the Coral Triangle Initiative on Coral Reefs, Fisheries, and Food Security which focuses on coastal and marine bio-resources' (VornDick 2015). The MPC-SCS workshops series is another reservoir of proposals such as a potential mechanism for regional fisheries cooperation, and a marine data information exchange and networking project that can be used (KEMLU 2021).

Conclusion

The social-ecological system in the SCS is in distress. This wicked problem has become more intractable over time – risks from territorialisation, militarisation, extraction and maritime traffic have accumulated to such an extent that environmental deterioration is all but guaranteed. This is compounded by global climate change. In its Sixth Assessment Report, the Intergovernmental Panel on Climate Change (IPCC) stated that the risk of species loss is greatest in the oceans, that warm-water corals are at the highest risk for decline, and that risks to coastal geographies increase with sea level rise.

Rising sea levels could, in turn, lead to some loss of territory, especially in the islands and islets in the South China Sea. The default reaction would be further construction over deteriorating areas. And would changing sea levels require states to redraw the geographical coordinates of their EEZs? If the land recedes and the EEZ coordinates are adjusted, that might mean more high seas. But this is an unknown variable as UNCLOS does not have any provisions for this eventuality (VornDick 2012). With some oil fields maturing, and with the agreement reached by the 28th Conference of Parties to the UN Framework Convention on Climate Change (COP 28) signalling the 'beginning of the end' of the fossil fuel era, there is speculation that there will be no economic reason to fight over resources in the South China Sea (Quiggin 2023).

Yet relying on environmental degradation and species loss as conditions for conflict resolution is wrongheaded, for four reasons. First, it ignores the 271 million people who will be directly affected by the South China Sea's deterioration. Second, while resource nationalism is hard to justify without the resources, nationalism is a sentiment that cannot simply be turned off. Third, militarisation increases possibilities of accidents and inadvertent use of force. Fourth, the most urgent task is to resuscitate the marine environment. This would require governance rather than alliance mechanisms, no matter how 'ironclad' the latter are proclaimed to be. The one certainty right now is that the environment will be further degraded without any ameliorative action and without steps towards transforming the current social subsystem.

Given these considerations, this article recommends that policymakers orient their governance away from principles of ownership and towards principles of stewardship (Liu 2017). A climate-centred code of conduct could build on, or modify, existing mechanisms such as 'climate hotlines', the MPC-SCS workshops and networks, or the Southeast Asian Fisheries Development Centre (SEAFDEC) for practical cooperation, augmenting the role of the Malacca Straits Patrol to include catching poachers in international waters, or adopting proposals presented in various regional forums. They could also develop 'communities of practice' by broadening ASEAN's environmental networks. A study of these networks found that they 'have formed practices that embody material resources, formal and informal institutions, and knowledge claims, as well as structures for collective meaning-making, and participation', but more work needs to be done to integrate climate change networks into decision-making (Salazar 2023, 203). Most importantly, Indigenous local knowledge and practices need to be seriously included, and communities must be empowered, deputising them to be a government's front lines.

Disclosure statement

No potential conflict of interest was reported by the author(s).

Acknowledgements

The author thanks the participants of the RSIS Roundtable 'Climate Security in the Indo-Pacific: Strategic Implications for Defense and Foreign Affairs' (2 November 2022) for their comments, and Cameron Olson for his research assistance on this project.

Notes

1. It is estimated that fish catch from IUUF was between 11 and 26 million tonnes per year, 'for an estimated value of US$10–23 billion', in the years leading up to 2014 (FAO 2016, 15–16).
2. These include the US-based Center for Climate and Security, the Center for Naval Analyses' (CNA) Military Advisory Board, the German-based Adelphi, the Stockholm International Peace Research Institute, the Dutch government-funded Planetary Security Initiative, the International Military Council on Climate and Security (IMCCS) and the Climate Security Expert Network.
3. B.L. Cherkasskii, "The System of the Epidemic Process," *Journal of Hygiene Epidemiology Microbiology and Immunology* 32, 3 (1988): 321, quoted in (Colding and Barthel 2019, 1). The term 'social-ecological system' was introduced in the 1970s but has become relevant because of the environmental damage wrought by human interaction and consumption (see Refulio-Coronado et al. 2021; Colding and Barthel 2019).
4. According to the United Nations Educational, Scientific and Cultural Organization (UNESCO), LMEs 'are regions of the world's oceans, encompassing coastal areas from river basins and estuaries to the seaward boundaries of continental shelves and the outer margins of the major ocean current systems'. The world's 66 LMEs 'produced about 80% of global annual marine fishery biomass … [and] contribute USD 12.6 trillion in goods and services each year to the global economy'.
5. Data from the US National Oceanic and Atmospheric Administration (NOAA); see https://www.lmehub.net/#south-china-sea.
6. The World Economic Forum considers 1945–1989 as 'Globalisation 2.0' when multinational companies exported their factories, and when world exports comprised between 5 and 15% of the world gross domestic product (GDP; Vanham 2019). Other scholars label this period as the 'great acceleration, when petroleum-powered globalization became the key driver in the transformation of the biosphere' (Dalby 2014, 4).
7. Viet Nam occupied 16 features, China had six, Malaysia had three, and the Philippines and Taiwan each occupied one feature.
8. Indonesia's and Malaysia's combined total oil production was around 1500 to 2000 Mbl/day during this period, but this number includes Indonesian oil blocks located outside of the South China Sea.
9. The Maritime Assertiveness Project is composed of a panel of international experts, and is supported by the National Bureau of Asian Research and the Sasakawa Peace Foundation. It looked at four types of assertiveness moves: declarative, demonstrative, coercive, and use of force. It also studied the issue areas involved, namely energy, fisheries, administrative control, military-security, and political support. The report by Chubb (2022) studied only China, Viet Nam and the Philippines' actions from 1970 to 2015.
10. Concerns about the warming effect of greenhouse gasses were made public by the 1987 Brundtland Report, and in a 1988 US Congressional testimony by NASA scientist James E. Hansen (Conway 2008).
11. WEF considers 1989–2008 as Globalisation 3.0, when trade accounted for half of the world's GDP (Vanham 2019).
12. By 2011, oil production from the South China Sea was estimated at 1.3 billion barrels/day and natural gas production was 3400 billion cu.ft.
13. According to Greer (2016), coral coverage rates in disputed areas of the SCS 'have declined from over 60 percent to just 20 percent' since the 1990s, and that 'virtually all SCS fishery stocks

are collapsed (roughly 25 percent), over-exploited (roughly 25 percent), or fully-exploited (roughly 50 percent)'. Between 1960 and 2000, highly popular species such as tuna, mackerel jacks, and sharks had been reduced by 50% (McManus et al. 2016).

14. According to the Burke et al. (2012), littoral countries of the SCS have high to very high social and economic vulnerability to reef loss.
15. Another source stated that between 2010 and 2016, Chinese maritime law enforcement vessels were involved in 79% of the 70 major incidents that occurred in the SCS (China Power Team 2016).
16. The term usually refers to distributional questions over a country's resources; but it is used here as 'a geopolitical discourse about sovereignty, the state, and territory, as well as the rights and privileges of citizenship, national identity, and the values a group assigns to resources' (Koch and Perreault 2019, 612).
17. The FAO (2014) estimates that between 1994 and 2014, IUUF catch went from 11 to 26 million tonnes with values reaching between $10 to $23 billion (FAO 2014).
18. This ranking is based on an Integrated Local Threat Index which includes 'threats from local activities of overfishing and destructive fishing, coastal development, watershed-based pollution, and marine-based pollution and damage'. See https://iuufishingindex.net/ranking.
19. The researchers used Visible Infrared Imaging Radiometer Suite (VIIRS) and synthetic aperture radar (SAR).
20. China is not the only 'guilty' party. The so-called blue boats of Viet Nam are criticised as criminal actors, and there are reports that Philippine law enforcement officials had disguised themselves as fishers in order to deter IUUF. It is also reported that some Philippine government agencies were enlisting some fishers to become a militia force, but Filipino fishermen from Subic and Zambales 'have not been very supportive of the idea' (Zhang and Bateman 2017, 292).

Bibliography

Abraham, M., J. Aguilar, J. Ballew, B. Cseke, B. Huneg, L. Latch, H. Mitchell, M. Nelson, S. Padmanabhan, and N. Wade. 2020. "Is Ecocide Inevitable in the World's Most Contested Sea?" *CSIS Journalism Bootcamp*, June 9, 2020. https://journalism.csis.org/is-ecocide-inevitable-in-the-worlds-most-contested-sea/

ADB. n.d. "The Coral Triangle: An Ecosystem under Threat." Asian Development Bank. Accessed February 22, 2023. http://www.adb.org/multimedia/coral-triangle.

Agnew, J. 1994. "The Territorial Trap: The Geographical Assumptions of International Relations Theory." *Review of International Political Economy* 1 (1): 53–80. https://doi.org/10.1080/09692299408434268.

AMTI. 2020. "Chinese Power Projection Capabilities in the South China Sea." Asia Maritime Transparency Initiative. https://amti.csis.org/chinese-power-projection/

AMTI. 2022. "Occupation and Island Building." Asia Maritime Transparency Initiative. https://amti.csis.org/island-tracker/

Baker, R. 2016. "Fish: The Overlooked Destabilizer in the South China Sea." *Stratfor*. February 12, 2016. https://worldview.stratfor.com/article/fish-overlooked-destabilizer-south-china-sea

Basu, P., and A. Chaturvedi. 2021. "In Deep Water: Current Threats to the Marine Ecology of the South China Sea." 449. ORF Issue Brief. New Delhi, India: Observer Research Foundation. On web. https://www.orfonline.org/research/in-deep-water-current-threats-to-the-marine-ecology-of-the-south-china-sea/.

Bateman, S. 2011. "Solving the 'Wicked Problems' of Maritime Security: Are Regional Forums up to the Task?" *Contemporary Southeast ASIA* 33 (1): 1–28. https://doi.org/10.1355/cs34-1a.

Blair, D. 2016. "U.S. Policy Options in the South China Sea – Senate Hearing 114-725." July 13. https://www.govinfo.gov/content/pkg/CHRG-114shrg27231/html/CHRG-114shrg27231.htm

Blecher, R. 2022. "Climate Change and Human Rights." International Crisis Group, July 28, 2022. https://www.crisisgroup.org/global/climate-change-and-human-rights.

Burke, M., S. M. Hsiang, and E. Miguel. 2015. "Climate and Conflict." *Annual Review of Economics* 7 (1): 577–617. https://doi.org/10.1146/annurev-economics-080614-115430.

Burke, L., K. Reytar, M. Spalding, and A. Perry. 2011. *Reefs at Risk Revisited*. Washington, D.C.: World Resources Institute.

Burke, L., K. Reytar, M. Spalding, and A. Perry. 2012. *Reefs at Risk Revisited in the Coral Triangle*. Washington, D.C.: World Resources Institute.

Busby, J. W. 2021. "Beyond Internal Conflict: The Emergent Practice of Climate Security." *Journal of Peace Research* 58 (1): 186–194. https://doi.org/10.1177/0022343320971019.

Bush, S. R., and M. Marschke. 2016. "Social and Political Ecology of Fisheries and Aquaculture in Southeast Asia." In *Routledge Handbook of the Environmental in Southeast Asia*, edited by P. Hirsch, 224–238. London: Routledge. https://edepot.wur.nl/413265.

Buzan, B., O. Wæver, and J. de Wilde. 1998. *Security: A New Framework for Analysis*. Boulder: Lynne Rienner.

Caballero–Anthony, M. 2017. "From Comprehensive Security to Regional Resilience: Coping with Nontraditional Security Challenges." In ASEAN @50 Volume 4: *Building ASEAN Community: Political-Security and Socio-Cultural Reflections*, edited by A. Baviera and L. Maramis, 123–145. Jakarta, Indonesia: Economic Research Institute for ASEAN and East Asia. https://www.eria.org/ASEAN_at_50_4A.7_Caballero-Anthony_final.pdf.

Chen, X., W. Deng, and W. E. I. Gangjian. 2023. "Coral Calcification History in the South China Sea since the Industrial Era and Its Response to Ocean Warming." *Quaternary Sciences* 43 (2): 448–457. https://doi.org/10.11928/j.issn.1001-7410.2023.02.13.

China Power Team. 2016. "Are Maritime Law Enforcement Forces Destabilizing Asia?" *ChinaPower Project*. August 18, 2016. https://chinapower.csis.org/maritime-forces-destabilizing-asia/

Chubb, A. 2022. "Dynamics of Assertiveness in the South China Sea: China, the Philippines, and Vietnam, 1970-2015." 99. NBR Special Report. Seattle, WA: The National Bureau of Asian Research.

Colding, J., and S. Barthel. 2019. "Exploring the Social-Ecological Systems Discourse 20 Years Later." *Ecology and Society* 24 (1). https://doi.org/10.5751/ES-10598-240102.

Conway, E. 2008. "NASA – What's in a Name? Global Warming vs. Climate Change." Feature Articles, Feature. NASA. Brian Dunbar. December 5, 2008. https://www.nasa.gov/topics/earth/features/climate_by_any_other_name.html

Corry, O. 2012. "'Securitisation and 'Riskification': Second-Order Security and the Politics of Climate Change." *Millennium: Journal of International Studies* 40 (2): 235–258. https://doi.org/10.1177/0305829811419444.

Cradock-Henry, N. A., J. Connolly, P. Blackett, and J. Lawrence. 2020. "Elaborating a Systems Methodology for Cascading Climate Change Impacts and Implications." *MethodsX* 7 (April):100893. https://doi.org/10.1016/j.mex.2020.100893.

Dalby, S. 2014. "Rethinking Geopolitics: Climate Security in the Anthropocene." *Global Policy* 5 (1): 1–9. https://doi.org/10.1111/1758-5899.12074.

Deudney, D. 1999. "Environmental Security: A Critique." In *Contested Grounds: Security and Conflict in the New Environmental Politics*, edited by D. Deudney and R. Matthews, 187–219. New York: SUNY Press.

Emmerson, D. K. 2020. *The Deer and the Dragon: Southeast Asia and China in the 21st Century*. Stanford: Shorenstein APARC/ISEAS – Yusof Ishak Institute.

Fabinyi, M., B. Belton, W. H. Dressler, M. Knudsen, D. S. Adhuri, A. Abdul Aziz, M. A. Akber, et al. 2022. "Coastal Transitions: Small-Scale Fisheries, Livelihoods, and Maritime Zone Developments in Southeast Asia." *Journal of Rural Studies* 91 (April): 184–194. https://doi.org/10.1016/j.jrurstud.2022.02.006.

FAO. 2014. *The State of World Fisheries and Aquaculture: Opportunities and Challenges*. Rome: Food & Agriculture Organization (FAO).

FAO. 2016. *FAO Yearbook: Fishery and Aquaculture Statistics*. Rome: FAO.

FAO. 2020. *The State of World Fisheries and Aquaculture 2020*. Rome: FAO. https://doi.org/10.4060/ca9229en.

FAO. 2022. "The State of World Fisheries and Agriculture 2022: Towards Blue Transformation." FAO. https://www.fao.org/publications/sofia/2022/en/

Ferrer, A. J. G., R. Pomeroy, M. J. Akester, U. Muawanah, W. Chumchuen, W. C. Lee, P. G. Hai, and K. Kuperan Viswanathan. 2021. "COVID-19 and Small-Scale Fisheries in Southeast Asia: Impacts and Responses." *Asian Fisheries Science* 34 (1): 99–113. https://doi.org/10.33997/j.afs.2021.34.1.011.

FitzGibbon, J., and K. O. Mensah. 2012. "Climate Change as a Wicked Problem: An Evaluation of the Institutional Context for Rural Water Management in Ghana." *SAGE Open* 2 (2): 215824401244848. https://doi.org/10.1177/2158244012448487.

Fleishman, R. 2022. "Climate Change, Security, and Political Coherence in the South and East China Seas: A Scenarios-Based Assessment." Prevention Web. 2022. https://www.preventionweb.net/publication/climate-change-security-and-political-coherence-south-and-east-china-seas-scenarios

Garba, S. I. 2022. "Wicked Problems: Threats and Opportunities." In In 9th Peace and Conflict Resolution Conference, Bangkok, Thailand.

Gaynor, J. L. 2005. "The Decline of Small-Scale Fishing and the Reorganization of Livelihood Practices among Sama People in Eastern Indonesia." *Michigan Discussions in Anthropology* 15 (1): 91–149.

Global Environment Facility. n.d. "Large Marine Ecosystems." https://www.thegef.org/what-we-do/topics/internationalwaters/marine/large-marine-ecosystems.

Green Climate Fund. n.d. "Green Climate Fund." Accessed February 24, 2023. https://www.greenclimate.fund/

Greer, A. 2016. "The South China Sea Is Really a Fishery Dispute." *The Diplomat*. July 20, 2016. https://thediplomat.com/2016/07/the-south-china-sea-is-really-a-fishery-dispute/

Guan, K. C., and J. K. Skogan, eds. 2007. *Maritime Security in Southeast Asia*. London: Routledge.

Head, B. W. 2022. *Wicked Problems in Public Policy: Understanding and Responding to Complex Challenges*. Cham, Switzerland: Palgrave Macmillan.

Heydarian, R. J. 2023. "US Cranking Pressure on China in South China Sea." *Asia Times*. April 3, 2023. https://asiatimes.com/2023/04/us-cranking-pressure-on-china-in-south-china-sea/

Ho, J., and S. Bateman, eds. 2012. *Maritime Challenges and Priorities in Asia: Implications for Regional Security*. London: Routledge.

Hodal, K., and J. Kaiman. 2014. "At Least 21 Dead in Vietnam Anti-China Protests over Oil Rig." *The Guardian*, May 15, 2014, sec. World news. https://www.theguardian.com/world/2014/may/15/vietnam-anti-china-protests-oil-rig-dead-injured

Hou, X., R. Li, and Z. Song. 2022. "A Bibliometric Analysis of Wicked Problems: From Single Discipline to Transdisciplinarity." *Fudan Journal of the Humanities and Social Sciences* 15 (3): 299–329. https://doi.org/10.1007/s40647-022-00346-w.

IISS. 2015. *Asia-Pacific Regional Security Assessment 2015: Key Developments and Trends*. London: International Institute for Strategic Studies.

IMCCS Expert Group. 2021. "World Climate and Security Report 2021." International Military Council on Climate and Security. https://imccs.org/wp-content/uploads/2021/06/World-Climate-and-Security-Report-2021.pdf

IMF Staff. 2008. "Globalization: A Brief Overview." International Monetary Fund. May 2008. https://www.imf.org/external/np/exr/ib/2008/053008.htm

International Energy Agency. 2023. "CO_2 Emissions in 2022." https://www.iea.org/reports/co2-emissions-in-2022

ISEAS. 2021. "The State of Southeast Asia: 2021 Survey Report." Singapore: ASEAN Studies Centre at ISEAS-Yusof Ishak Institute. https://www.iseas.edu.sg/wp-content/uploads/2021/01/The-State-of-SEA-2021-v2.pdf

Kalkavan, B. 2017. "Climate Change, Inequality and Conflict: Approaches to Strengthen Mali's Climate Change Adaptation." Clingandael Policy Brief. The Hague: Clingandael Netherlands Institute of International Relations.

KEMLU. 2021. *The Thirtieth Workshop on Managing Potential Conflict in the South China Sea*. Jakarta, Indonesia: Foreign Ministry of Indonesia.

Koch, N., and T. Perreault. 2019. "Resource Nationalism." *Progress in Human Geography* 43 (4): 611–631. https://doi.org/10.1177/0309132518781497.

Liu, N. 2017. "Towards a Code of Conduct for the South China Sea – Policy Forum." *Policy Forum*, September 8, 2017. https://www.policyforum.net/towards-code-conduct-south-china-sea/

Liu, Y., S. Tao, R. Sun, C.-F. You, T.-H. Wang, T. Felis, W. Zheng, et al. 2023. "Coral Growth over the past 550 Years in the Central South China Sea Linked to Monsoon- and Seabird-Induced Nutrient Stress." *Palaeogeography, Palaeoclimatology, Palaeoecology* 617 (May): 111488. https://doi.org/10.1016/j.palaeo.2023.111488.

Mahadzir, D. 2022. "RIMPAC 2022 Kicks Off in Hawaii with 21 Partner Nation Ships." *USNI News*, June 29, 2022. https://news.usni.org/2022/06/29/rimpac-2022-kicks-off-in-hawaii-with-21-partner-nation-ships

McClatchy Newspapers. 2009. "US Study Warns of Pollution from Merchant Ships off Florida Coast." *The Guardian*, March 31, 2009, sec. Environment. https://www.theguardian.com/environment/2009/mar/31/noaa-pollution-florida-freighters-tankers-cruise-ships

McDonald, M. 2013. "Discourses of Climate Security." *Political Geography* 33: 42–51. https://doi.org/10.1016/j.polgeo.2013.01.002.

McManus, J., E. Gomez, S. Wells, S. Norman, and S. Jupiter. 2016. "Coral Reefs of the South China Sea – A Need for Action." International Society for Reef Studies. https://coralreefs.org/wp-content/uploads/2019/01/SCS_Briefing_ISRS_20161711_revised_v3.pdf

McPhail, P., A. Speyer, B. Rodgers, S. Ostrosky, J. Burns, and D. Marquis. 2020. "The Strategic Littoral Geography of Southeast Asia." Center for International Maritime Security. June 4, 2020. https://cimsec.org/the-strategic-littoral-geography-of-southeast-asia/

NDTV World News. 2022. "South China Sea Level Has Risen By 150 Mm Since 1900: Study." NDTV. March 28, 2022. https://www.ndtv.com/world-news/south-china-sea-level-has-risen-by-150-mm-since-1900-study-2847248.

Oceans Inc. 2021. "Fishers on the Frontlines." Oceans Inc. August 23, 2021. https://www.oceansinc.earth/fishers-on-the-frontlines

Ortuoste, M. 2018. "Strategic and 'Everyday' Maritime Security Challenges in Southeast Asia." In *Naval Powers in the Indian Ocean and the Western Pacific*, edited by H. Hensel and A. Gupta, 120–137. London: Routledge.

Pauly, D., D. Zeller, and M. L. D. Palomares, eds. 2020. "Sea Around Us Concepts Design and Data." https://www.seaaroundus.org/

Poling, G. B. 2019. "Illuminating the South China Sea's Dark Fishing Fleets." Stephenson Ocean Security Project. January 9, 2019. https://ocean.csis.org/spotlights/illuminating-the-south-china-seas-dark-fishing-fleets/

Quiggin, J. 2023. "Climate Change, Ironically, Reduces the Heat in the South China Sea." The Interpreter. March 30, 2023. https://www.lowyinstitute.org/the-interpreter/climate-change-ironically-reduces-heat-south-china-sea

Refulio-Coronado, S., K. Lacasse, T. Dalton, A. Humphries, S. Basu, H. Uchida, and E. Uchida. 2021. "Coastal and Marine Socio-Ecological Systems: A Systematic Review of the Literature." *Frontiers in Marine Science* 8: 1–17. https://doi.org/10.3389/fmars.2021.648006.

Rittel, H. W. J., and M. M. Webber. 1973. "Dilemmas in a General Theory of Planning." *Policy Sciences* 4 (2): 155–169. https://doi.org/10.1007/BF01405730.

Ruttinger, L., R. Munayer, P. van Ackern, and F. Titze. 2022. "The Nature of Conflict and Peace: The Links between Environment, Security and Peace and Their Importance for the United Nations." Gland, Switzerland: WWF International and Adelphi Consult. https://www.adelphi.de/en/publication/nature-conflict-and-peace-links-between-environment-security-and-peace-and-their

Salazar, V. 2023. "From Practices to Praxis: ASEAN's Transnational Climate Governance Networks as Communities of Practice." *Journal of Current Southeast Asian Affairs* 42 (2): 190–215. https://doi.org/10.1177/18681034231167443.

Sekiyama, T. 2022. "Climate Security and Its Implications for East Asia." *Climate* 10 (7): 104. https://doi.org/10.3390/cli10070104.

Singapore Maritime and Port Authority. 2023. "Vessel Information Traffic System." https://mpa.gov.sg/port-marine-ops/operations/vessel-traffic-information-system

SIPRI. 2022. "World Military Expenditure Passes $2 Trillion for First Time." SIPRI. April 25, 2022. https://www.sipri.org/media/press-release/2022/world-military-expenditure-passes-2-trillion-first-time

Smith, L., P. Cornillon, D. Rudnickas, and C. B. Mouw. 2019. "Evidence of Environmental Changes Caused by Chinese Island-Building." *Scientific Reports* 9 (1): 5295. https://doi.org/10.1038/s41598-019-41659-3.

Solomon, M. 2016. "RIMPAC Invites Ships, World Nations – and Environmental Fury." *Hawai'i Public Radio*, June 30, 2016. https://www.hawaiipublicradio.org/general-assignment/2016-06-30/rimpac-invites-ships-world-nations-and-environmental-fury

Storey, I. 2017. "Anatomy of the Code of Conduct Framework for the South China Sea." The National Bureau of Asian Research (NBR). August 24, 2017. https://www.nbr.org/publication/anatomy-of-the-code-of-conduct-framework-for-the-south-china-sea/

Sumaila, U. R., and W. W. L. Cheung. 2016. "Boom or Bust: The Future of Fish in the South China Sea." Canada. https://www.oceanrecov.org/news/ocean-recovery-alliance-news/boom-or-bust-the-future-of-fish-in-the-south-china-sea.html

Sumaila, U. R. 2019. "Comparative Valuation of Fisheries in Asian Large Marine Ecosystems with Emphasis on the East China Sea and the South China Sea LMEs." *Deep Sea Research Part II: Topical Studies in Oceanography* 163: 96–101. https://doi.org/10.1016/j.dsr2.2018.12.008.

Teh, L. C. L., and D. Pauly. 2018. "Who Brings in the Fish? The Relative Contribution of Small-Scale and Industrial Fisheries to Food Security in Southeast Asia." *Frontiers in Marine Science* 5: 1–9. https://doi.org/10.3389/fmars.2018.00044.

Thayer, C. 2018. "A Closer Look at the ASEAN-China Single Draft South China Sea Code of Conduct." *The Diplomat*, August 3, 2018. https://thediplomat.com/2018/08/a-closer-look-at-the-asean-china-single-draft-south-china-sea-code-of-conduct/

Theisen, O. M. 2017. "Climate Change and Violence: Insights from Political Science." *Current Climate Change Reports* 3 (4): 210–221. https://doi.org/10.1007/s40641-017-0079-5.

Turnbull, N., and R. Hoppe. 2019. "Problematizing 'Wickedness': A Critique of the Wicked Problems Concept, from Philosophy to Practice." *Policy and Society* 38 (2): 315–337. https://doi.org/10.1080/14494035.2018.1488796.

US Energy Information Agency. 2024. "South China Sea." https://www.eia.gov/international/analysis/regions-of-interest/South_China_Sea

Vanham, P. 2019. "A Brief History of Globalization." World Economic Forum. January 17, 2019. https://www.weforum.org/agenda/2019/01/how-globalization-4-0-fits-into-the-history-of-globalization/

Vergun, D. 2023. "New EDCA Sites Named in the Philippines." U.S. Department of Defense. April 3, 2023. https://www.defense.gov/News/News-Stories/Article/Article/3350297/new-edca-sites-named-in-the-philippines/https%3A//www.defense.gov/News/News-Stories/Article/Article/3350297/new-edca-sites-named-in-the-philippines/

VornDick, W. 2012. "Thanks Climate Change: Sea-Level Rise Could End South China Sea Spat." *The Diplomat*, November 8, 2012. https://thediplomat.com/2012/11/can-climate-change-wash-away-south-china-sea-dispute/

VornDick, W. 2015. "Confidence Building through Climate Change in the South China Sea." *Asia Maritime Transparency Initiative*, December 4, 2015. https://amti.csis.org/confidence-building-through-climate-change-in-the-south-china-sea/

Wæver, O. 1995. "Securitization and Desecuritization." In *On Security*, edited by R. D. Lipschutz, 46–86. New York: Columbia University Press.

Wirth, C. 2018. *Danger, Development and Legitimacy in East Asian Maritime Politics: Securing the Seas, Securing the State*. London: Routledge.

Xiao, J., W. Wang, X. Wang, P. Tian, and W. Niu. 2022. "Recent Deterioration of Coral Reefs in the South China Sea Due to Multiple Disturbances." *PeerJ* 10 (July): e13634. https://doi.org/10.7717/peerj.13634.

Xinhua. 2021. "Chinese Premier Urges Expedited Negotiations of the COC." *Xinhuanet*, October 26, 2021. http://www.news.cn/english/2021-10/26/c_1310270430.htm

Xu, K. 2009. "Piracy and Energy Security in Southeast Asian Waters." In *Asian Energy Security: The Maritime Dimension*, edited by H. Lai, 183–203. New York: Palgrave Macmillan.

Yuan, X., Y. Guo, W.-j. Cai, H. Huang, W. Zhou, and S. Liu. 2019. "Coral Responses to Ocean Warming and Acidification: Implications for Future Distribution of Coral Reefs in the South China Sea." *Marine Pollution Bulletin* 138 (January): 241–248. https://doi.org/10.1016/j.marpolbul.2018.11.053.

Zhang, H., and S. Bateman. 2017. "Fishing Militia, the Securitization of Fishery and the South China Sea Dispute." *Contemporary Southeast Asia* 39 (2): 288–314. https://doi.org/10.1355/cs39-2b.

Zhang, T., Y. Qu, Q. Zhang, J. Tang, R. Cao, Z. Dong, Q. Wang, and J. Zhao. 2021. "Risks to the Stability of Coral Reefs in the South China Sea: An Integrated Biomarker Approach to Assess the Physiological Responses of Trochus Niloticus to Ocean Acidification and Warming." *Science of the Total Environment* 782 (August): 146876. https://doi.org/10.1016/j.scitotenv.2021.146876.

OPEN ACCESS

Shifting discourses of climate security in India: domestic and international dimensions

Dhanasree Jayaram

ABSTRACT

The Indian perspectives on climate security are influenced by both domestic and international imperatives. The logic followed by India is not typically the same as that adopted by countries of the Global North. India's discourses on the interconnections between climate change and security are largely conditioned by developmental priorities (domestic) and geopolitical pressures (international), which are not necessarily mutually exclusive. In this context, the article contextualises India's climate security narratives at the intersection of domestic and international/foreign-policy realms, involving different actors, including academia and think tanks, the government and its agencies, the military and non-governmental organisations. It argues that there has been a gradual process of integrating climate change into the domestic security policy-making agenda in the country, with the growing recognition of worsening climate vulnerabilities, while demonstrating reluctance to engage with the existing climate security discourses at the international level. It analyses the ways in which the 'domestic' and 'international' intermingle with each other to shape India's discourses and practices on climate security by categorising them into two dimensions – domestic–international and international–domestic.

Introduction

The widespread acknowledgement of climate change as an international, national and human security issue has led to the incorporation of climate change in many countries' foreign-policy-making and security strategies, especially that of member countries of the Organisation for Economic Co-operation and Development. Climate security (or security implications of climate change) is usually linked with a host of threats/risks including the disrupting effects of climate change on societies and economies, conflicts over scarce or degraded resources, population displacement and mobility patterns related to climate-related disasters, and erosion of resilience of communities (Odeyemi 2020). Climate security has also become an integral part of the agenda of several international and regional organisations such as the United Nations Security Council (UNSC), European Union (EU), and North Atlantic Treaty Organisation (Vogler 2023). Within the UN system, the UN Climate Security Mechanism (UNCSM) was established in 2018 as a joint initiative between the United Nations Development Programme (UNDP), UN Department of Political and Peacebuilding Affairs, United Nations Environment Programme, later joined by the UN Department of Peace Operations 'to strengthen the capacity of the United Nations to more systematically analyze and address the linkages between climate change, peace and security' (UNCSM 2022). The recognition of the security implications of climate change is seen not only as a response to the mounting body of evidence linking security and climate change, but also as a tactic to infuse a sense of urgency into global climate governance and accelerate climate action at the international, national and local levels.

In India, while the discourse(s) on climate security at the domestic level may be considered limited, there is an increasing acknowledgement of human and national security implications of climate change by various actors, such as academia and think tanks, the government and its agencies, and the military, among others. Many of these conceptualisations are however not reflected in the policy process adequately, as India has not fully formalised its security strategy (except in the case of its military doctrine) and when formal security discussions happen, they are usually dominated by conventional security challenges (Jayaram 2021a). Additionally, there has not been much research on the connections between climate change and security, aside from some peer-reviewed literature (Barthwal-Datta 2012; Sahu 2019) and grey literature produced by think tanks (IPCS 2022; IDSA 2011). What is also noteworthy is that India is one of the countries that, up until 2023, have opposed any efforts to give the UNSC the authority to intervene in any climate change-related decision-making processes, implying its reservations about the climate security narratives in the international milieu. This includes voting against the first draft resolution on climate security in December 2021, when India was a non-permanent member in the council (Mitra 2021).

In this context, the article first provides an analytical framework through which the research puzzle is disentangled, examining the shifts in India's international and domestic discourses on climate security – analysing how and why these distinct domestic and international dimensions interact with each other. The first section refers to literature that explains interactions between the foreign and domestic policy-making in the field of International Relations. This framework would help analyse the reasons for India's resistance to acknowledge climate security at the international level, despite increasing domestic concerns about climate change as well as the growing influence of international climate security discourses on India's domestic narratives on the issue, thereby offering a relational analysis of India's foreign and domestic policies on climate security. It then takes a deep-dive into the international and domestic policy discourses on climate security, involving the above-mentioned actors. Thereafter, it expounds the two dimensions of India's climate security discourses – focussing upon the influence of India's domestic factors on its international climate security discourses, and that of the international factors on domestic climate security discourses.

What also needs to be noted here is that in the absence of overt demonstrations of a climate security agenda in India, these linkages are built largely though secondary and grey literature (meant for policy consumption and at times involving knowledge co-production with policy communities), interviews with policy and security communities, and indirect references to climate change and security (such as through water security, food security, energy security, etc.) that can be found in policy/official documents. Hence, the article also provides an overview of policy research (conducted primarily by the country's security policy community), institutions, critical issues and other factors that have an influence on India's climate security discourses.

When it comes to India's engagement with international platforms on climate security, the article primarily focusses on the UNSC, but also incorporates examples from other forums such as the United Nations Framework Convention on Climate Change (UNFCCC) and regional organisations in South Asia and the broader Indo-Pacific. The UNSC, since 2007, has assumed the lead in attempting to address climate-related security risks through open debates, high-level debates and 'arria-formula' meetings, albeit with limited policy impact (Maertens and Trombetta 2023). These regular debates have been sponsored mostly by industrialised countries, but also increasingly supported by least developed countries and island states. However, UNSC resolutions on climate security have been consistently blocked by Russia, a permanent (veto-wielding) member (Hardt 2021).

The UNFCCC is the apex UN entity tasked with the responsibility of tackling climate change. The recognition of climate-conflict linkages and other human security implications of climate change by the Intergovernmental Panel on Climate Change's (IPCC) has led to greater emphasis on these issues in the UNFCCC discussions, particularly in relation to refugee movement, human well-being and disasters – with references to conflicts in countries such as Sudan, Syria, and Ukraine (UNFCCC 2022). However, climate security is not yet institutionalised as a part of the UNFCCC's mandate as also seen by its absence in the UNCSM. The UNCSM, on the other hand, is a more of a knowledge-producing and capacity-building platform that brings together analysts and practitioners to build capacities within the UN to address climate-related security risks, and not a decision-making and/or enforcement body (UNCSM 2022).

'Domestic' and 'international' realms in climate security discourses

The literature on foreign-policy-making has long engaged with notions of how domestic politics influences foreign policy and vice versa. Within this literature, the two-level game theoretic model is one of the oft-used frameworks for analysing the interplay between the domestic and international dynamics (Putnam 1988). While much of the literature has focussed upon domestic determinants of foreign policy, there is a significant body of work that has also looked into the effects of the international system on domestic politics, which is referred to as 'the second image reversed' by Gourevitch (1978). According to Gourevitch (1978, 882–883), 'the distribution of power among the states' and the 'distribution of economic activity and wealth' (apart from 'ideas') influence domestic regimes and policy-making.

The two-level game theoretic model has been used in multiple studies to analyse the positions of different states. For instance, international bargaining processes of parties such as the EU, the United States (US) and Russia have been linked with domestic constraints, including the politics of ratification (Lisowski 2002; Mukhia 2018; Milkoreit 2019). Recent literature concerning India's climate policy/diplomacy has also focussed on the role of ideas and interests in shaping it by intertwining domestic and international factors – without necessarily engaging with the two-level game model (Dubash et al. 2018; Sengupta 2019; Jayaram 2021b). For instance, Dubash et al. (2018) highlight internalisation of India's international climate pledges and the impacts of India's discourses on integration of development with climate change at the domestic level on its international policies.

When it comes to climate security, some studies analyse the international and domestic discourses on climate security of specific countries, but largely treat them as separate realms of decision-making. For instance, developed countries like the United Kingdom (UK) tend to externalise climate security, combining traditional and human security concerns, and emphasising 'adaptation and resilience, particularly for countries and regions deemed acutely vulnerable to climatic and political upheaval' (Harrington 2023, 295). On the other hand, for most developing countries, the domestic framings of climate security are influenced by development imperatives with greater emphasis on poverty reduction, equitable access to resources (energy, water, etc.), socio-economic well-being, including by providing employment opportunities and protecting the most vulnerable populations, etc. (Ouweneel and Simpson 2023).

However, the lack of engagement with international–domestic interactions in the existing climate security literature eludes a holistic understanding of climate security approaches espoused by states, especially when it comes to exploring questions of why states like India adopt a position at one level despite having contrarian discourses at the other level. Similarly, the existing literature focusses more on the Global North and tends to externalise climate security as a problem of the Global South, wherein the agency of domestic and foreign-policy-making processes of countries in the Global South is overlooked. The climate security discourses of the US, UK, Germany, France and many other countries of the Global North typically reflect this notion, whereby conflicts or migration associated with climate change in the Global South affect security and stability of the Global North (Von Lucke 2020; Harrington 2023; Estève 2023). Hence, there is also a need to delve into the contextual meanings and practices of climate security in the Global South. The purpose of this article is to locate and analyse the evolution of India's (a developing country) positions on climate security at the international level vis-à-vis the domestic discourses. It provides a relational analysis of India's positions on climate security at the international and domestic levels; and analyses the influence of domestic factors on its foreign policy and that of the international factors on its domestic policy.

The other aspect of climate security that needs to be highlighted further is the historicity attached to these discourses that are influenced by colonial legacies and development inequities that are often overlooked in the climate security literature. A deeper engagement with postcolonial critiques of both the climate change and security discourses offers a reflexive representation and conceptualisation of climate security discourses in many countries. Herein, a lens that does not necessarily see national, human and other formulations of security as distinct or mutually exclusive assumes significance. Therefore, this article will also contribute to a pluralistic understanding of climate security from a Global South perspective that is currently limited.

Indian perspectives on international climate security discourses

India's positions on the existing climate security narratives globally have been a mixed bag. It has remained opposed to discussing security implications of climate change in the UNSC, one of the international platforms that has been at the forefront of discussing climate-related security risks/threats. In 2021, when a draft resolution to acknowledge climate security risks was introduced, the Indian representative to the UN, T. S. Tirumurti (UNSC 2021), asserted:

> India is second to none when it comes to climate action and climate justice. But the UN Security Council is not the place to discuss either issue. In fact, the attempt to do so appears to be motivated by a desire to evade responsibility in the appropriate forum and divert the world's attention from an unwillingness to deliver where it counts...To move forward decisively, affordable access to climate finance and technologies has become critical. Developed countries must provide climate finance of $1 trillion at the earliest. It is necessary that climate finance be tracked with the same diligence as climate mitigation. And the reality ... is that the developed countries have fallen well short of their promises. This is particularly important to recognize because today's attempt to link climate with security really seeks to obfuscate lack of progress on critical issues under the UNFCCC process. This draft resolution would only sow the seeds of discord among the larger UN membership. It sends a wrong message to the developing countries that instead of addressing their concerns and holding developed countries responsible for meeting their commitments under the UNFCCC, we are willing to be divided and side-tracked under the guise of security. This draft resolution is a step backward from our collective resolve to combat climate change. It seeks to hand over that responsibility to a body which neither works through consensus nor is reflective of the interests of the developing countries. India had no option but to vote against.

The Indian position in the UNSC is guided by several factors, including the belief that the global nature of the climate change issue diverges with 'security' that is seen as being constrained by territorial boundaries, thereby rendering securitisation of climate change an unviable solution (Chaudhury 2020). Another argument given by the Indian representatives belonging to the Ministry of External Affairs (MEA) is the apparent lack of or the 'contested' nature of evidence for climate-conflict nexus (through causal mechanisms) (UNSC 2019). For instance, in the 2007 open discussion on climate security in the UNSC, the Indian representative (diplomat) argued that poverty, energy competition and a lack of development—rather than climate change—were what caused conflicts (UNSC 2007).

Importantly, for India, the UNSC is an unrepresentative forum that is not fit for discussing climate change, considering it has only 15 member states with five permanent ones possessing veto power (Jayaram 2023b). Ruchira Kamboj, India's first woman Permanent Representative to the UN stated in 2022 that India's opposition to discussing climate security in the UNSC was 'based on principles' (PTI 2022). India contends that the UNFCCC and other more representative UN agencies should deal with climate change, thereby also emphasising the need for 'a broader approach, anchored in development, adaptive capacity, risk assessment and institutional build-up', rather than a securitised or militarised approach (UNSC 2011; UNSC 2019). For instance, there are concerns over the risk of sanctions against non-performing countries, aimed at coercing them into adopting more climate commitments than they are prepared to undertake domestically (Scott 2015). This also raises questions over whether such coercive measures would be applied only to the less powerful, non-veto countries, while veto-wielding ones (that are also the world's biggest greenhouse gas emitters) might still get away.

Concerns of coercion are also closely connected with India's scepticism regarding principles such as Responsibility to Protect (R2P), which implies (humanitarian) intervention, indicative of militarisation (Abdenur and Folly 2020). India argued against invoking R2P in the case of the 2008 Cyclone Nargis in Myanmar, while many Western countries leaned towards it when the Myanmar regime refused entry to foreign aid workers and relief material in the aftermath of the cyclone (Singh 2020). India has expressed concerns about the effects of R2P on nations' 'sovereignty', the UNSC's involvement in interventions (especially when it remains unreformed), and certain states' unilateral actions such as sanctions (Aneja 2014; Singh 2020).

For India, the UNFCCC plays an influential part in defining its climate security discourses at the international level (Jayaram 2023b). For example, one cannot overlook a general mistrust of the industrialised countries' intentions (stemming from the North-South tensions). The industrialised countries' 'historical responsibility' and the principle of Common but Differentiated Responsibilities and Respective Capabilities (CBDR-RC) are inflexible for India (Jayaram 2021b; Sengupta 2019). India's statements in the earlier UNSC discussions indicate a sense of concern that financial and technological resources (overseas development assistance) from rich nations intended for development projects and poverty eradication could be diverted towards climate mitigation (UNSC 2007). Although this sentiment has considerably softened with greater support for climate action domestically over time, scepticism over climate security-related discussions in international organisations is palpable. In the UNFCCC too, India has consistently stressed on 'additionality', whereby 'funds raised for climate change should not substitute or divert funds from other important developmental objectives, particularly social and economic development' (Mandal 2019, 382).

The politics of climate change is at times linked with the strategic timing of the 2007 UNSC meeting, the first time that climate change was introduced in the UNSC. Thereafter, the UN General Assembly (UNGA), a more representative forum (at least when it comes to the system of 'one country, one vote'), unanimously adopted a resolution on 'Climate change and its possible security implications' (UNGA 2009a). It released a report of the Secretary-General later in the same year, reinforcing the 2007 meeting's climate security agenda, even while recognising the UNFCCC as the apex climate change-related decision-making body (UNGA 2009b). These developments raise questions about the intentions of the industrialised countries, especially amidst the 2007–2009 financial crisis that reduced even further their willingness to contribute to climate finance required by developing countries (Jalles 2023). The 2007 meeting took place just ahead of the Bali climate summit, wherein the strict differentiation between the developed and developing countries (as operationalised in the Kyoto Protocol) was somewhat diluted, culminating in the failure of the 2009 Copenhagen Summit to reach an agreement (Jayaram 2021b). Hence, this could have been seen by India as a strategic move by the industrialised countries to force countries such as India to adopt legally binding emissions reduction targets (Jayaram 2023b).[1]

Apart from the global narratives, regional dynamics also have an impact on India's discourses. To start with, narratives of inter-state/cross-border 'climate migration' and 'climate refugees' exemplify traditional notions of security, as in the political discourse, 'illegal immigration' from Bangladesh to India is connected to climate change-related disasters and resource competition, with socio-economic and political repercussions for India (Chowdhary 2020; Bose 2014). Even though most policy experts assert that climate change is unlikely to cause new conflicts, they also reiterate that it may exacerbate existing insecurities, thereby adopting the threat multiplier logic (IDSA 2011; Dasgupta 2016; Saran 2019; IPCS 2022). For instance, variations in river water levels (induced by climate change) could affect transboundary river water management with Pakistan and China over the sharing of the Indus and the Brahmaputra, respectively, with whom India shares geopolitically hostile relations (Klare 2020; Sahu and Mohan 2022; Jayaram and Sethi 2022b). Indian think tanks, while shaping climate security narratives, have also identified regional issues such as glacial recession, sea level rise, and extreme weather events that could spark off different scenarios, including 'civil wars', 'military invasion', 'economic migration' and 'political refugees' (Pai 2017). Similarly, in the maritime realm – Indian Ocean Region – climate security risks such as displacement and migration, maritime disputes (owing to climate impacts on low-lying islands), and potential conflicts over 'Exclusive Economic Zones (EEZs) and seabed resources' have been discussed by serving and ex-military officials (Singh 2015).

Climate security is also contextualised in South Asia and the broader Indo-Pacific region through the lens of cooperation. For instance, as a strategy of addressing climate change-related disasters, India coordinates with its neighbours and nations in the Indo-Pacific region, especially through Humanitarian Assistance and Disaster Relief (HADR) operations as well as pre-disaster HADR exercises aimed at strengthening disaster response capabilities. These efforts are integral to India's soft power, tied to India's aspiration to be seen as a net security provider in the region and a responsible power globally (The Print 2022; Gong and Jayaram 2023). Similarly, India has also been cooperating with the other countries of the region through regional organisations such as the South Asian Association for Regional Cooperation, the Bay of Bengal Initiative for Multi-Sectoral Technical and Economic Cooperation, the Association of Southeast Asian Nations, and the Indian Ocean Naval Symposium (IONS) – all of which identify climate change as an area of cooperation (Jayaram 2022a). India engages with security groupings such as the Quad (involving the US, Japan, Australia, and India) closely. The Quad recognises risks posed by climate change to peace and stability in the Indo-Pacific region, and has launched the 'Quad Climate Change Adaptation and Mitigation Package' (Q-CHAMP) and finalised the guidelines for 'Quad Partnership on Humanitarian Assistance and Disaster Relief (HADR) in the Indo-Pacific' (The White House 2021; The White House 2022).

Domestic discourses on climate security in India

Domestically, India's perceptions of climate security are not necessarily tied to the framings of climate security at the international level, barring a few exceptions. The discourses on development heavily influence the Indian perspectives on climate security, which contrasts it with the Western and/or Global North schools of thinking on climate security. The recognition of security implications of climate change plays out mostly through an assessment of the threats posed by climate change to the country's economic development. If one draws upon the securitisation framework of the Copenhagen School, it could be argued that India has securitised climate change by emphasising these threats and adopting 'emergency' policy and institutional measures such as enacting the 2008 National Action Plan on Climate Change (NAPCC) and establishing the Prime Minister's Council on Climate Change (PMCCC) in late 2000s (Sahu 2019). Similar arguments have been made by Barthwal-Datta (2012), according to whom climate change is securitised in India by non-state actors like 'scientific policy communities' that act as 'securitising actors' apart from the state. Organisations such as The Energy and Resources Institute and the Centre for Science and Environment have been instrumental in preparing the NAPCC, thereby being involved in the securitisation process on account of their 'social capital' and 'expert authority' (Barthwal-Datta 2012).

India's policies at the intersection of climate change and security draw upon water security, food security, energy security and livelihood security, which are linked with development issues. The underlying causes of harm or insecurity are taken into account by merging various issues. For instance, India has prioritised food security since the Green Revolution was launched in 1960s, including by enacting the National Food Security Act, 2013, which focusses more on socio-economic aspects of access and affordability (DFPD 2013). A certain constitutive logic, coupled with India's agrarian identity, which links the issue with economic security dimensions as well (Chattopadhyay 2010; Das and Ansari 2021), brings out the ways in which the risk of erosion of developmental gains caused by climate change is emphasised. For instance, Chauhan (2009, 3) highlights the threats posed by climate change-related disasters to poverty reduction, development and economic growth.

Similarly, on water security, increasing scarcity to the point where certain big cities could potentially experience a 'day zero' disaster (when taps run completely dry) has also underscored the need for incorporating climate considerations into water security-related policies across the country (Suneja 2019). India's 2012 National Water Policy mentions 'the possibility of deepening water conflicts among different user groups' over water owing to constraints posed by the growing population and energy demand, coupled with climate change (MWR 2012). In 2018–19, the metropolitan city of Chennai faced a 'day zero' situation, when its taps ran dry and people had to stand in queues for hours to collect water from public and private tankers. It also resulted in conflicts between urban and rural water users (Guntoju, Faiz Alam, and Sikka 2019). Similarly, river water sharing conflicts among the Indian states (for example, the Cauvery/Kaveri dispute between Karnataka and Tamil Nadu) could be affected by climate change as it exacerbates water scarcity concerns in combination with other factors such as urbanisation (Sharma, Hipel, and Schweizer 2020).

On the other hand, many critical voices have resisted linking climate change with security or emergency domestically. This stems from the interlinkages built between development and climate security in terms of the negative impacts of climate mitigation on development priorities. Ramesh (2015), former Minister of State (MoS, Independent Charge), Ministry of Environment and Forests (now Ministry of Environment, Forest and Climate Change or MoEFCC), stated that the ministry is often branded 'anti-people' as it is seen to be obstructing economic progress for ecological reasons. According to Ramesh (2015), neither ecological security nor development (and employment opportunities) can be guaranteed at the expense of one another. This sentiment has been consistently espoused by all the administrations. India's current MoS, MoEFCC, Bhupender Yadav, states, 'India's climate policy is directed towards sustainable development and poverty eradication while striving continuously to decouple emissions from growth and achieve energy efficiency across sectors' (PIB 2023). In practice, unless it is related to the sustainable development agenda, climate action (especially mitigation) is generally seen as partially damaging to the country's developmental priorities (Dubash et al. 2018). At the policy level, these narratives are used to rationalise building capacities and acquiring financial and technological resources through economic growth and development as the only pathway towards stronger climate action, rather than securitising climate change (Dasgupta 2013).

India's perspectives of climate security in the domestic (non-governmental) realm are guided by a distinct Indian environmental philosophy, steered by 'environmentalism of the poor.' Herein, justice is a crucial consideration, whether it is about conflicts over natural resource management or survival itself. The ecological imperative is one among several other goals of environmental conservation (Shiva 2009) and climate action. For instance, energy transition, aimed at reducing greenhouse gas emissions, would also help achieve energy security, thereby combining the 'co-benefits' of both climate and energy security

(Bisht 2012; Dubash 2013; Ramesh 2015). In addition, the security implications of energy transition (such as conflicts over land use and impacts on livelihoods) are also increasingly being discussed, thereby pushing for a just transition agenda, which is yet again being steered by both state and non-state actors, such as think tanks and research institutions (Banerjee 2021).

In the absence of an official (publicly available) national security strategy, the Indian establishment's views on security are generally captured in publications produced by, or statements made by, political, bureaucratic, and non-governmental policy/academic communities (who work closely with the government in an advisory capacity). For instance, in India's National Security Annual Review 2015–2016, the late Chandrasekhar Dasgupta, one of India's most experienced negotiators in the UNFCCC processes and a former member of the PMCCC, highlighted climate change as a non-traditional security challenge and 'threat multiplier' (Dasgupta 2016). This review is published by the Foundation for National Security Research, a think tank in New Delhi and typically contains essays by experts working on various aspects of 'national security,' including members of the National Security Council (NSC) and/or National Security Advisory Board (NSAB), who provide advice to the Prime Minister's Office (PMO). However, since 2018, the think tank has not published the annual reviews, and the PMCCC has not met since 2015, after it was reconstituted by the newly elected government in 2014 (Sangomala 2021). Furthermore, 'Resource, Water and Food Security' was formerly on the NSAB's agenda (Bolton 2017), particularly between 2010 and 2015, and it examined the effects of climate change on national security, primarily in terms of resource availability. Devendra Kumar Sharma, member of the NSAB (since 2021) and Chairman, Himachal Pradesh Electricity Regulatory Commission, has looked into the effects of climate change on transboundary rivers such as the Indus (with the view of revising the Indus Waters Treaty signed between India and Pakistan), among other issues concerning water management, in his articles (Sharma 2023). This lack of continuity and consistency renders the task of analysing India's climate security perspectives difficult.

The 2017 Joint Doctrine of the Indian Armed Forces labels climate change as a 'non-traditional security' issue, with potential implications for migration/displacement, disaster response, civil strife etc., thereby necessitating 'security' responses (HQ-IDS 2017). This in a way reflects the changing perceptions of climate change by the Ministry of Defence (MoD) through a security lens. Several serving and ex-military personnel have also contributed to climate security discourses at the policy level. For instance, Lieutenant General (retired) Vinod G. Khandare (an ex-Indian Army officer), Military Advisor in the NSC Secretariat under the PMO (2018–2021) and Principal Adviser in the MoD (since 2022), emphasises the need to consider climate and broader environmental issues in threat assessment matrix and security planning. He highlights threats posed by weather modification and similar geoengineering techniques, climate migration and other transboundary climate security concerns to India's security[2]. Similarly, a report on climate security in India published by the Institute of Peace and Conflict Studies (IPCS), an Indian think tank, in partnership with the Clingendael Institute, a Dutch think tank, based on input provided by governmental and non-governmental actors (including military ones), states, 'Climate change effects can create the conditions for conflict by aggravating intermediate 'stressors', whether they are social, political, and/or economic' (IPCS 2022).[3] Hence, the conflict potential of climate change, albeit not endorsed in formal policy process, has become a part of the climate discourses in the country.

The Indian military is increasingly becoming aware of climate security risks for its operations and strategies. Chauhan highlights the threats posed by climate risks such as rising sea level, cyclones and storm-surges to the Indian Navy's naval bases and installations. He also brings out the indirect impacts of climate change on 'the ability of the navy to carry out its operations owing to rising air and ocean temperatures, changes in salinity, as also oceanic oxygen depletion and the increase in bioluminescent algae blooms, ocean acidification and changes in ambient underwater noise, etc.'[4] Similarly, Air Marshal (retired) Anil Chopra explains the impact of high temperatures on load carriage for aircraft and sensitive avionics as well as underscores the contribution of the armed forces to global greenhouse gas emissions (Chopra 2023).

A relational analysis of India's discourses on climate security

Domestic–international dimension

This section further substantiates the influence of domestic factors on India's international positioning on climate security. In the UNSC and other international forums that discuss climate security, or even broadly climate change, India has consistently drawn upon climate justice and North-South discourses, such as during its G-20 presidency in 2023 (Saxena 2023; Hurrell and Sengupta 2012). For instance, Chaturvedi and Doyle (2015, 77), being critical of the existing Western climate security discourses, observe that the notions of climate security used in the West do not reflect the realities of the Global South that has faced the brunt of climate change. They argue that the Western discourses do not even distinguish between luxury emissions and survival emissions, where the former are linked to developed countries and the latter to developing ones (Chaturvedi and Doyle 2015, 153). Boas (2014) points out that rather than focussing on issues of national or international security, the Indian perspective on climate change and security is informed by ideas of 'human security' and 'energy security'. The 'alarmist' language on climate security, including on topics like climate migration, has long been resisted by India, which has viewed this discourse as 'arrogant' (Boas 2014) and a strategy employed by the Global North to pressure India to adopt emissions reduction obligations within the UNFCCC.

One of India's major concerns is that 'securitisation' of climate change could lead to further vilification of coal, on which India's energy security and development is highly dependent. 55% of India's energy needs are still met by coal (MoC 2023). In such a scenario, as also exemplified by India's move to advocate for 'phase-down' rather than 'phase-out' of coal at the 2021 Glasgow climate summit (Mohan 2021), India is apprehensive of additional pressure to phase out coal owing to involvement of international organisations such as the UNSC that has a history of playing interventionist politics (Chaudhury 2020). India is not yet prepared to abandon coal completely despite appreciating the additional benefits of climate action for energy security through renewable energy deployment (Kanitkar, Mythri, and Jayaraman 2022). Clearly, the domestic development imperatives influence India's positioning on climate security on international platforms such as the UNSC.

Linking climate security with mitigation commitments has been a dilemma for India, considering its position on mitigation commitments has been contingent on the need to maintain autonomy, which is to keep domestic climate action independent of international dynamics and the pressure to act more (Saran 2008; Jayaram 2021b). In the 2007 open debate in the UNSC, the Indian representative categorically said that developing countries' emissions posed no threat to global peace and security, and that it was the responsibility of wealthier nations to lower their own (UNSC 2007). In theory, this is a position taken by all the developing and least developed countries. However, compared to the majority of other developing, least developed and island countries for which these vulnerabilities are at the centre of their climate policy, which also invariably has an impact on their perceptions of security itself, India's immense physical vulnerability to climate change has not significantly influenced its positions, either in the climate change negotiations or in the UNSC (Jayaram 2021b). India's international climate policy has mostly focussed upon a low-carbon development strategy with the intention of being recognised as a 'responsible power' rather than an obstacle in international climate cooperation, especially within the UNFCCC processes (Sengupta 2019).

International–domestic dimension

In this section, the influence of international factors on domestic discourses and policies on climate change are discussed in more detail. Yet again, transnational climate justice and North–South disparities in terms of both development and climate vulnerabilities are critical to this dimension. For instance, the concept of 'environmental colonialism' used by Agarwal and Narain (2019) to describe the climate change negotiations supports a longstanding mistrust of Western-centric discourses on the subject. They highlight that these discourses lean towards 'one worldism' and 'high minded internationalism' in the name of 'morality'. The postcolonial worldviews, with their impact on the development and environmental paradigms espoused by countries such as India, are at the centre of climate security narratives too. For India, in the post-independence era, development was a pathway towards maintaining sovereignty (protect the country from foreign interference), building a modern identity (drawing upon Western notions of modernity and industrial development), and lifting millions out of people of poverty (attributed to the colonial experience) (Sirohi 2019, 67–111). This Global North-driven development paradigm, while contributing to these efforts, has also been responsible for socio-ecological destruction in various places (Chatterjee 2019).

In fact, 'historicisation' of environmental security (Zwierlein 2017) and postcolonial interpretations of 'ecological security' (Jayaram 2023a) could provide a more nuanced understanding of the prevailing climate security discourses both in the Global North and Global South. One needs to move beyond the post-second world war and post-cold war narratives of growing ecological consciousness and redefining/broadening security respectively. There is also a need to problematise the evolution of meanings of 'nature' and 'security' that have been heavily influenced by the dominant ideologies of 'imperial economies' and concepts of 'nation state' that were hallmarks of the colonial era (Zwierlein 2017). Many Indian scholars such as Swami (2003), Gadgil and Guha (2004), Prakash (2006), Rajan (2006), Gupta (2009) and Saravanan (2023) have provided extensive accounts of large-scale socio-ecological destruction caused by the colonial (British) empire in India, thereby calling for separating the Indian variant of environmentalism from the 'post-industrial' and 'post-material' Western interpretations of environmentalism (Gadgil and Guha 2004).

In spite of these efforts to develop counter-narratives on environmental and climate change issues, concepts such as climate-resilient development (often used in the climate security literature) and low-carbon development (specifically targeted at climate mitigation) have received greater traction in India as climate vulnerabilities grow and the willingness to engage with and adopt the normative approaches of global climate governance (Sengupta 2019; Singh and Chudasama 2021). For instance, a report on climate change, peace and security, published by Kubernein Initiative, a geopolitical advisory/think tank in India, in partnership with adelphi, a Berlin-based think tank/consultancy, recommends community and infrastructural resilience in climate security practices in India and broader South Asia (KI and adelphi 2024).[5] With greater international pressure to commit to emissions reduction and to be seen as a responsible power, the political and public acceptance for these norms and approaches has grown domestically. Yet, although major international organisations embrace the language of climate security, resilience and similar concepts, there are still gaps on the domestic front in India, when it comes to diffusion of these norms, considering the complexity of issues that operate at multiple scales in a non-linear and systemic manner (Unnikrishnan and Nagendra 2021). Similarly, despite the growing rhetorical acceptance of 'threat multiplier' and 'risk multiplier' discourses, their domestic policy impact remains negligible (Jayaram 2021a; Dasgupta 2016).

Conclusion

Within India's security discourses and architectures, climate security remains underconceptualised. Yet, there is a growing acknowledgement of the interlinkages between climate change and security. Although Western/Global North narratives such as the 'threat multiplier' logic and conflict-centric framings influence some of the domestic discourses on climate security, the larger objective has been to distance from the Western narratives and outline a climate security narrative that could be contextualised in the Indian case, as stressed by think tanks such as the IPCS. At the same time, India's positions in various multilateral forums such as the UNSC goes beyond the conceptualisations of climate security to reflect upon the geopolitical complexities involved in addressing climate security.

The relational analytical (international and domestic) approach used in this article explains the various ways in which India's perspectives on climate security are constructed through different processes, facilitated by several actors (and not just the state). While India's positions on climate security internationally are not entirely reflective of its domestic discourses, in the absence of a formal security strategy, it is also difficult to scrutinise its domestic policies on climate change through a security lens. Climate security remains a non-institutionalised realm in the Indian context owing to the contradictions within the discourses climate security as well as India's own balancing act between development and climate action.

However, it is also evident that India's inability to redefine security in a systematic manner, despite increasingly invoking non-traditional security framings to define issues such as climate change at the domestic level, has served as a deterrent to a clearly identifiable climate security policy domestically. Nevertheless, with newly emerging geopolitical dynamics in the Indo-Pacific region and India's burgeoning regional engagement on climate change (including through a security lens), India could be pushed further to integrate climate security into its foreign-policy approaches. It is also important to acknowledge that the security framings depend on the actors. Even if certain non-governmental actors and the military (and to a lesser extent the MoD) may be seen as more willing to examine climate security, many government actors such as the Ministry of Coal (MoC), MoEFCC and MEA (and increasingly the PMO) do not seem to be on the same page.

To sum up, the relational analysis reveals the interactions between the international and domestic discourses of climate security espoused by India – providing the rationale for varied positions on the issue at the domestic and international levels. As much as the constraining domestic factors lead to its reluctant international positioning on the issue (by the state), one could still discern the diffusion of international climate security discourses into the Indian domestic ones, channelised through different actors (beyond the state) and shaped by domestic imperatives. Through a strong empirical focus, describing and contextualising the case of India, this article has explained the nuances of the construction of climate security discourses and practices in the Global South, which have often been overlooked by global discourses that are tilted towards Western notions of climate security.

Disclosure statement

The author declares that she has no conflict of interest.

Acknowledgements

The author expresses deepest gratitude to the guest editors of the special issue on 'Climate Security in the Indo-Pacific: Variations, Contestations and Convergence of Security Practices' – Prof. Mely Caballero-Anthony and Dr Alistair D. B. Cook, S. Rajaratnam School of International Studies (RSIS), Nanyang Technological University, Singapore – for giving her an opportunity to contribute an article to this issue. She is also grateful to them and their team at RSIS for organising a Roundtable on 'Climate Security in the Indo-Pacific: Strategic Implications for Defense and Foreign Affairs', on November 2, 2022, in which she received constructive comments on a preliminary draft of this article. Special thanks to Dr Florian Krampe (Stockholm International Peace Research Institute, Sweden), who, as a discussant in this roundtable, provided valuable feedback on the draft. She would also like to thank the two reviewers for their incisive remarks that helped her revise and fine-tune the article; and the editors of Third World Quarterly for their continuous support.

Notes

1. Personal communication with a think tank expert and journalist who has followed the climate change negotiations and UNSC discussions on climate security since 2000s.
2. Lt. Gen. (retired) Vinod Khandare made these remarks in his distinguished lecture on 'Comprehensive National Power and India's National Security', held on 19 January 2024, at Manipal Academy of Higher Education, Karnataka, India.
3. The author was a member of the task force (2021–2022) constituted by the IPCS to assess and find ways to address the security implications of climate change for South Asia. The task force comprised 'Indian academics, climate experts, scholar-activists, technical experts, and security practitioners'. For more information, please visit: https://www.planetarysecurityinitiative.org/sites/default/files/2022-12/IPCSSR%20214_Six%20Questions%20on%20CS%20for%20India_Nov%202022.pdf.
4. Personal communication with Vice Admiral (retired) Pradeep Chauhan, Director General, National Maritime Foundation, New Delhi, India, dated 18 September 2021.
5. The author participated in the Roundtable Workshop on 'Population, Physiography, Precarity – Collectively Building a Climate for Peace in South Asia', organised by Kubernein Initiative in partnership with adelphi, in New Delhi, during 7–8 November 2023. The workshop was attended by 'a diverse range of actors from government, civil society, think tanks, and academia

from South Asia and beyond'. For more information, please visit: https://kuberneininitiative.com/article/roundtable-discussion-on-population-physiography-precarity-collectively-building-a-climate-for-peace-in-south-asia-i-7th-and-8th-november-2023/.

ORCID

Dhanasree Jayaram http://orcid.org/0000-0003-0483-7943

References

Abdenur, A. E., and M. Folly. 2020. "The United Nations and Anthropocenic Governance." *Strategic Analysis* 44 (5): 463–468. https://doi.org/10.1080/09700161.2020.1825063

Agarwal, A., and S. Narain. 2019. "Global Warming in an Unequal World: A Case of Environmental Colonialism." In *India in a Warming World: Integrating Climate Change and Development*, edited by N. K. Dubash, 81–91. New Delhi: Oxford Academic.

Aneja, U. 2014. "India, R2P and Humanitarian Assistance." *Global Responsibility to Protect* 6 (2): 227–245. https://doi.org/10.1163/1875984X-00602008

Banerjee, S. 2021. "Defining and Framing Just Transition for India." India Just Transition Center, Working Paper no. 1. https://iforest.global/wp-content/uploads/2021/11/Working-paper-1.pdf

Barthwal-Datta, M. 2012. *Understanding Security Practices in South Asia: Securitization Theory and the Role of Non-State Actors*. Abingdon: Routledge.

Bisht, T. C. 2012. "Energy Security and Climate Change Challenges: India's Dilemma and Policy Responses." In *Energy Security in the Era of Climate Change. Energy*, edited by L. Anceschi and J. Symons, 111–125. London: Palgrave Macmillan.

Boas, I. 2014. "Where Is the South in Security Discourse on Climate Change? An Analysis of India." *Critical Studies on Security* 2 (2): 148–161. https://doi.org/10.1080/21624887.2014.905295

Bolton, L. 2017. "National Security Office Responsibilities and Functions." *K4D Research Helpdesk*, February 27, 2017. https://assets.publishing.service.gov.uk/media/5980885ded915d022b000040/K4D_HDR__National_Security_Advisor_Office.pdf

Bose, S. 2014. "Illegal Migration in the Indian Sunderbans." *Forced Migration Review* 45: 22. https://www.fmreview.org/sites/fmr/files/FMRdownloads/en/crisis/bose.pdf

Chatterjee, E. 2019. "The Asian Anthropocene: Electricity and Fossil Developmentalism." *The Journal of Asian Studies* 79 (1): 3–24. https://doi.org/10.1017/S0021911819000573

Chattopadhyay, N. 2010. "Climate Change and Food Security in India." In *Climate Change and Food Security in South Asia*, edited by R. Lal, M. V. K. Sivakumar, S. M. A. Faiz, A. H. M. M. Rahman, and K. R. Islam, 229–250. Dordrecht: Springer.

Chaturvedi, S., and T. Doyle. 2015. *Climate Terror: A Critical Geopolitics of Climate Change*. Basingstoke: Palgrave Macmillan.

Chaudhury, D. R. 2020. "India Warns against Securitisation of Environmental Issues." *The Economic Times*, September 18, 2020. https://economictimes.indiatimes.com/news/politics-and-nation/india-warns-against-securitisation-of-environmental-issues/articleshow/78182043.cms?from=mdr

Chauhan, S. 2009. "Climate Change, Disasters and Security… Issues, Concerns and Implications for India." *United Services Institute*. https://www.semanticscholar.org/paper/Climate-Change-%2C-Disasters-and-Security-…-…-%2C-Chauhan/d746c8a3dd405b2c3d4228dfd24f49fb36c27a25

Chopra, A. 2023. "Why Indian Armed Forces Must Institutionalise Climate Change in Their Planning and Operations." *First Post*, November 2, 2023. https://www.firstpost.com/opinion/why-indian-armed-forces-must-institutionalise-climate-change-in-their-planning-and-operations-13333442.html

Chowdhary, N. 2020. "Environmental Refugees: A Humanitarian Crisis in India and Bangladesh." In *Refugee Crises and Third-World Economies*, edited by S. K. Das and N. Chowdhary, 37–43. Bingley: Emerald Publishing Limited.

Das, U., and M. A. Ansari. 2021. "The Nexus of Climate Change, Sustainable Agriculture and Farm Livelihood: Contextualizing Climate Smart Agriculture." *Climate Research* 84: 23–40. https://doi.org/10.3354/cr01648

Dasgupta, C. 2016. "Climate Change and India's Security Environment." In *India's National Security Annual Review 2015–16*, edited by S. Kumar, 389–398. New Delhi: Routledge.

Dasgupta, C., Jr. 2013. "Keynote Address." In *The U.S.-India Strategic Relationship: Cross Sector Collaboration to Promote Sustainable Development*, edited by M. J. Fratantuono, D. M. Sarcone, and J. Colwell, Carlisle: Dickinson College and United States Army War College.

DFPD. 2013. "National Food Security Act (NFSA)." Department of Food and Public Distribution, Ministry of Consumer Affairs, Government of India. Accessed August 20, 2023. https://dfpd.gov.in/nfsa.htm

Dubash, N. 2013. "The Politics of Climate Change in India: Narratives of Equity and Cobenefits." *WIREs Climate Change* 4 (3):191–201. https://doi.org/10.1002/wcc.210

Dubash, N. K., R. Khosla, U. Kelkar, and S. Lele. 2018. "India and Climate Change: Evolving Ideas and Increasing Policy Engagement." *Annual Review of Environment and Resources* 43 (1): 395–424. https://doi.org/10.1146/annurev-environ-102017-025809

Estève, A. 2023. "Preventing and Managing Climate Risks: France's Approach to Climate Security." In *Climate Security in the Anthropocene*, edited by J. N. Hardt, C. Harrington, F. Von Lucke, A. Estève, and N. P. Simpson, Vol. 33, 113–130. Cham: Springer International Publishing. https://doi.org/10.1007/978-3-031-26014-8_6

Gadgil, M., and R. Guha. 2004. *The Use and Abuse of Nature: Incorporating This Fissured Land & an Ecological History of India and Ecology and Equity*. New Delhi: Oxford University Press.

Gong, L., and D. Jayaram. 2023. "Status-Seeking through Disaster Relief Cooperation: China and India in Southeast Asia." In *Contemporary Southeast Asia* 45 (2): 246–281.

Gourevitch, P. 1978. "The Second Image Reversed: The International Sources of Domestic Politics." *International Organization* 32 (4): 881–912. https://doi.org/10.1017/S002081830003201X

Guntoju, S. S., M. Faiz Alam, and A. Sikka. 2019. "Chennai Water Crisis: A Wake-Up Call for Indian Cities." *Down to Earth*, August 5, 2019. https://www.downtoearth.org.in/blog/water/chennai-water-crisis-a-wake-up-call-for-indian-cities-66024

Gupta, S. D. 2009. "Accessing Nature: Agrarian Change, Forest Laws and Their Impact on an Adivasi Economy in Colonial India." *Conservation and Society* 7 (4): 227–238. http://www.jstor.org/stable/26392982. https://doi.org/10.4103/0972-4923.65170

Hardt, J. 2021. "The United Nations Security Council at the Forefront of (Climate) Change? Confusion, Stalemate, Ignorance." *Politics and Governance* 9 (4): 5–15. https://doi.org/10.17645/pag.v9i4.4573

Harrington, C. 2023. "Climate Change as a "Threat Multiplier": The Construction of Climate Security by the United Kingdom—2007–2020." In *Climate Security in the Anthropocene*, edited by J. N. Hardt, C. Harrington, F. von Lucke, A. Estève, and N. P. Simpson, Vol. 33, 297–318. Cham: Springer.

HQ-IDS. 2017. *Joint Doctrine Indian Armed Forces*. Headquarters Integrated Defence Staff, Ministry of Defence, Government of India. New Delhi. https://bharatshakti.in/wp-content/uploads/2015/09/Joint_Doctrine_Indian_Armed_Forces.pdf

Hurrell, A., and S. Sengupta. 2012. "Emerging Powers, North-South Relations and Global Climate Politics." *International Affairs* 88 (3): 463–484. https://doi.org/10.1111/j.1468-2346.2012.01084.x

IDSA. 2011. *Security Implications of Climate Change for India*. New Delhi: Manohar Parrikar Institute for Defence Studies and Analyses.

IPCS. 2022. "Task Force Report: Six Questions on Climate Security for India." *Institute of Peace & Conflict Studies*. https://www.planetarysecurityinitiative.org/news/task-force-report-six-questions-climate-security-india

Jalles, J. T. 2023. "Financial Crises and Climate Change." *Comparative Economic Studies*. https://doi.org/10.1057/s41294-023-00209-7

Jayaram, D. 2021a. "Climatizing' Military Strategy? A Case Study of the Indian Armed Forces." *International Politics* 58 (4): 619–639. https://doi.org/10.1057/s41311-020-00247-3

Jayaram, D. 2021b. *Climate Diplomacy and Emerging Economies: India as a Case Study*. Abingdon: Routledge.

Jayaram, D. 2022a. "The Politics of Climate Change in South Asia." In *Routledge Handbook of the International Relations of South Asia*, edited by S. Ganguly and F. O'Donnell, 411–424. London: Routledge.

Jayaram, D., and G. Sethi. 2022b. "Geopolitics of Climate Change and Water Security in South Asia: Conflict and Cooperation." In *Science, Policies and Conflicts of Climate Change*, edited by N. Khare, 77–88. Cham: Springer. https://doi.org/10.1007/978-3-031-16254-1_4

Jayaram, D. 2023a. "A Global South Perspective on 'Ecological Security.'" *New Perspectives* 31 (1): 31–38. https://doi.org/10.1177/2336825X221143624

Jayaram, D. 2023b. "Deconstructing Indian Discourses and Practices on Climate Change and Security." In *Handbook on Climate Change and International Security*, edited by M. J. Trombetta, 159–181. Cheltenham: Edward Elgar Publishing. https://doi.org/10.4337/9781789906448.00018

Kanitkar, T., A. Mythri, and T. Jayaraman. 2022. "Equity Assessment of Global Mitigation Pathways in the IPCC Sixth Assessment Report." *OSF Preprints*. November 3. https://doi.org/10.31219/osf.io/p46ty

KI and adelphi. 2024. "Population, Physiography, Precarity – Collectively Building a Climate for Peace in South Asia." Workshop Report. *Kubernein Initiative*, January 2024.

Klare, M. 2020. "Climate Change, Water Scarcity, and the Potential for Interstate Conflict in South Asia." *Journal of Strategic Security* 13 (4): 109–122. https://doi.org/10.5038/1944-0472.13.4.1826

Lisowski, M. 2002. "Playing the Two-Level Game: US President Bush's Decision to Repudiate the Kyoto Protocol." *Environmental Politics* 11 (4): 101–119. https://doi.org/10.1080/714000641

Maertens, L., and M. J. Trombetta. 2023. "Climate Change at the United Nations Security Council: Securitization, Climatization and Beyond." In *Handbook on Climate Change and International Security*, edited by M. J. Trombetta, 182–200. Cheltenham: Edward Elgar Publishing. https://doi.org/10.4337/9781789906448.00019

Mandal, K. 2019. "Climate Finance." In *India in a Warming World: Integrating Climate Change and Development*, edited by N. K. Dubash, 381–398. New Delhi: Oxford University Press.

Milkoreit, M. 2019. "The Paris Agreement on Climate Change—Made in USA?" *Perspectives on Politics* 17 (4): 1019–1037. https://doi.org/10.1017/S1537592719000951

Mitra, D. 2021. "In India's Negative Vote in UNSC, a Long-Standing Advocacy against Overlapping Mandates." *The Wire*, December 14, 2021. https://thewire.in/diplomacy/india-negative-vote-unsc-climate-change-overlapping-mandates.

MoC. 2023. "Coal – India's Energy Choice." Ministry of Coal, Government of India. https://coal.nic.in/en/major-statistics/coal-indian-energy-choice.

Mohan, G. 2021. "India Did Not Replace 'Phase Out' of Coal with 'Phase Down' in Glasgow Climate Pact, Criticism Unfair: Govt Sources." *India Today*, November 18, 2021. https://www.indiatoday.in/india/story/india-did-not-replace-phase-out-coal-phase-down-climate-pact-glasgow-govt-1877940-2021-11-17

Mukhia, A. 2018. "China's Two-Level Game in the Climate Change Negotiation." In *China's Foreign Relations and Security Dimensions*, edited by G. Kochhar, 145–162. London: Routledge.

MWR. 2012. "National Water Policy." Ministry of Water Resources, Government of India. https://nwm.gov.in/sites/default/files/national%20water%20policy%202012_0.pdf

Odeyemi, C. 2020. "The UNFCCC, the EU, and the UNSC: A Research Agenda Proposal for the Climate Security Question." *Advances in Climate Change Research* 11 (4): 442–452. https://doi.org/10.1016/j.accre.2020.11.004

Ouweneel, B., and N. P. Simpson. 2023. "Securing a Climate-Resilient Pathway for South Africa." In *Climate Security in the Anthropocene*, edited by J. N. Hardt, C. Harrington, F. von Lucke, A. Estève, and N. P. Simpson, Vol. 33, 231–257. Cham: Springer.

Pai, N. 2017. "Climate Change and National Security." *Centre for Land Warfare Studies*. https://takshashila.org.in/wp-content/uploads/2017/02/ClimateChange-NationalSecurity.pdf

PIB. 2023. "Shri Bhupender Yadav Says India's Climate Policy Is Directed towards Sustainable Development and Poverty Eradication While Striving Continuously to Decouple Emissions from Growth and Achieve Energy Efficiency across Sector." Press Information Bureau, Government of India. https://www.pib.gov.in/PressReleseDetailm.aspx?PRID=1904361

Prakash, O. 2006. "Wildlife Destruction: A Legacy of the Colonial State in India." *Proceedings of the Indian History Congress* 67: 692–702. http://www.jstor.org/stable/44147988.

PTI. 2022. "During UNSC Tenure, There Were Instances When India Had to Stand Alone: UN Envoy Ruchira Kamboj." *The Hindu*, December 23, 2022. https://www.thehindu.com/news/national/during-unsc-tenure-there-were-instances-when-india-had-to-stand-alone-un-envoy-ruchira-kamboj/article66295652.ece.

Putnam, D. R. 1988. "Diplomacy and Domestic Politics: The Logic of Two-Level Games." *International Organization* 42 (3): 427–460. https://doi.org/10.1017/S0020818300027697

Rajan, S. R. 2006. *Modernizing Nature: Forestry and Imperial Eco-Development, 1800–1950*. New York: Oxford University Press.

Ramesh, J. 2015. *Green Signals: Ecology, Growth, and Democracy in India*. New Delhi: Oxford University Press.

Sahu, A. K. 2019. "The Democratic Securitization of Climate Change in India." *Asian Politics & Policy* 11 (3): 438–460. https://doi.org/10.1111/aspp.12481

Sahu, A. K., and S. Mohan. 2022. "From Securitization to Security Complex: Climate Change, Water Security and the India–China Relations." *International Politics* 59 (2): 320–345. https://doi.org/10.1057/s41311-021-00313-4

Sangomala, A. 2021. "PM's Climate Change Council has not Met in Almost 7 Years." *Down to Earth*, November 25, 2021. https://www.downtoearth.org.in/news/climate-change/pm-s-climate-change-council-has-not-met-in-almost-7-years-80369.

Saran, S. 2008. "Climate Change – From Back Room to Board Room – What Indian Business Needs to Know about India's Approach to Multilateral Negotiations on Climate Change." Ministry of External Affairs, Government of India, April 21, 2008. https://mea.gov.in/in-focus-article.htm?18821/Talk±by±Special±Envoy±of±Prime±Minister±Shri±Shyam±Saran±in±Mumbai±on±Climate±Change

Saran, S. 2019. "Need for a Comprehensive National Security Strategy." *Policy Challenges – 2019-2024: The Big Policy Questions for the New Government and Possible Pathways*. Centre for Policy Research, June 4, 2019. New Delhi. https://cprindia.org/need-for-a-comprehensive-national-security-strategy/

Saravanan, V. 2023. *Colonialism and wildlife: An environmental history of modern India*. London: Routledge.

Saxena, N. 2023. "India's Climate Leadership Shines in New Delhi Declaration at G20 Summit." *The Times of India*, September 10, 2023. https://timesofindia.indiatimes.com/blogs/the-write-wing/indias-climate-leadership-shines-in-new-delhi-declaration-at-g20-summit/

Scott, S. V. 2015. "Implications of Climate Change for the UN Security Council: Mapping the Range of Potential Policy Responses." *International Affairs* 91 (6): 1317–1333. https://doi.org/10.1111/1468-2346.12455

Sengupta, S. 2019. "India's Engagement in Global Climate Negotiations from Rio to Paris." In *India in a Warming World: Integrating Climate Change and Development*, edited by N. K. Dubash, 114–141. New Delhi: Oxford University Press.

Sharma, A., K. W. Hipel, and V. Schweizer. 2020. "Strategic Insights into the Cauvery River Dispute in India." *Sustainability* 12 (4): 1286. https://doi.org/10.3390/su12041286

Sharma, D. K. 2023. "Indus Waters Treaty – Overtaken by Technology and Climate Change." Natstrat, September 2023. https://natstrat.org/articledetail/publications/indus-waters-treaty-overtaken-by-technology-and-climate-change-77.html

Shiva, V. 2009. "Water Wars in India." In *Globalisation and Environmental Challenges: Reconceptualising Security in the 21st Century*, edited by H. G. Brauch, Ú. O. Spring, C. Mesjasz, J. Grin, P. Dunay, N. C. Behera, B. Chourou, P. Kameri-Mbote, and P. H. Liotta, 589–592. Berlin: Springer.

Singh, A. 2015. "Climate Change and Maritime Security in the Indian Ocean Region." *Journal of Defence Studies* 9 (1): 63–82. http://idsa.in/jds/9_1_2015_ClimateChangeandMaritimeSecurityin.html.

Singh, A. 2020. "Indian Perspectives on the 'Responsibility to Protect." *International Studies* 57 (3): 296–316. https://doi.org/10.1177/0020881720930605

Singh, P. K., and H. Chudasama. 2021. "Pathways for Climate Resilient Development: Human Well-Being within a Safe and Just Space in the 21st Century." *Global Environmental Change* 68:102277. https://doi.org/10.1016/j.gloenvcha.2021.102277

Sirohi, R. A. 2019. *From Developmentalism to Neoliberalism: A Comparative Analysis of Brazil and India*. Singapore: Palgrave Macmillan.

Suneja, V. 2019. "Prioritise and Invest on Water Security and Clean Drinking Water." *Down to Earth*, December 27, 2019. https://www.downtoearth.org.in/blog/climate-change/prioritise-and-invest-on-water-security-and-clean-drinking-water-68591

Swami, V. 2003. "Environmental History and British Colonialism in India: A Prime Political Agenda." *CR: The New Centennial Review* 3 (3): 113–130. https://doi.org/10.1353/ncr.2004.0011

The Print. 2022. "India Has Emerged as Net Security Provider in Indo-Pacific: Defence Minister Rajnath Singh." November 29, 2022. https://theprint.in/world/india-has-emerged-as-net-security-provider-in-indo-pacific-defence-minister-rajnath-singh/1241982/

The White House. 2021. "Joint Statement from Quad Leaders." September 24, 2021. https://www.whitehouse.gov/briefing-room/statements-releases/2021/09/24/joint-statement-from-quad-leaders/

The White House. 2022. "Guidelines for Quad Partnership on Humanitarian Assistance and Disaster Relief (HADR) in the Indo-Pacific." September 23, 2022. https://www.whitehouse.gov/briefing-room/statements-releases/2021/09/24/joint-statement-from-quad-leaders/

UNFCCC. 2022. "Conflict and Climate." *United Nations Climate Change*, July 12, 2022. https://unfccc.int/news/conflict-and-climate.

UNGA. 2009a. "Climate Change and Its Possible Security Implications." United Nations General Assembly, June 11, 2009. https://www.securitycouncilreport.org/atf/cf/%7B65BFCF9B-6D27-4E9C-8CD3-CF6E4FF96FF9%7D/res%2063%20281.pdf

UNGA. 2009b. "Climate Change and Its Possible Security Implications: Report of the Secretary-General." United Nations General Assembly, September 11, 2009. https://digitallibrary.un.org/record/667264/files/A_64_350-EN.pdf?ln=en

Unnikrishnan, H., and H. Nagendra. 2021. "Building Climate Resilient Cities in the Global South: Assessing City Adaptation Plans in India." *The Round Table* 110 (5): 575–586. https://doi.org/10.1080/00358533.2021.1985268

UNSC. 2007. "Security Council Holds First-Ever Debate on Impact of Climate Change on Peace, Security, Hearing over 50 Speakers." United Nations Security Council, April, 17, 2007. https://www.un.org/press/en/2007/sc9000.doc.htm

UNSC. 2011. "Security Council, in Statement, Says 'Contextual Information' on Possible Security Implications of Climate Change Important when Climate Impacts Drive Conflict." United Nations Security Council, July 20, 2011. http://www.un.org/News/Press/docs/2011/sc10332.doc.htm

UNSC. 2019. "Massive Displacement, Greater Competition for Scarce Resources Cited as Major Risks in Security Council Debate on Climate-Related Threats." United Nations Security Council, January 25, 2019. https://www.un.org/press/en/2019/sc13677.doc.htm

UNSC. 2021. "Action on Draft Resolution on Security in the Context of Climate Change", Permanent Mission of India to the United Nations, December 13, 2021. https://www.pminewyork.gov.in/IndiaatUNSC?id=NDQ1NQ

Vogler, A. 2023. "Tracking Climate Securitization: Framings of Climate Security by Civil and Defense Ministries." *International Studies Review* 25 (2):viad010. https://doi.org/10.1093/isr/viad010

Von Lucke, F. 2020. *The Securitisation of Climate Change and the Governmentalisation of Security. New Security Challenges*. Cham: Springer International Publishing. https://doi.org/10.1007/978-3-030-50906-4

Zwierlein, C. 2017. "Historicizing Environmental Security." *European Journal for Security Research* 3 (1): 1–13. https://doi.org/10.1007/s41125-017-0026-4

Climate security and Japan's new national security strategy: a policy analysis

Hideshi Tokuchi

ABSTRACT
Japan's National Security Strategy makes it clear that the Government of Japan is more climate-security minded than before, although its overall tone is dominated by geopolitical and traditional security issues. The Government of Japan, and the Ministry of Defense in particular, should address climate security particularly in close cooperation with its international partners. Military-to-military cooperation should be accelerated mainly on the response and adaptation of the military forces to the impacts of climate change. Japan's effort will contribute to the international efforts to build a new security architecture in the Indo-Pacific. The Ministry of Defense established its Response Strategy on Climate Change. This is important as it clearly shows the Ministry's recognition that the effort to address climate change is compatible with national security efforts, including military efforts. As climate security is a truly global agenda, Japan's international security cooperation to address climate change should be pursued globally. As it is quite new for Japan, there is much room for expansion in response, adaptation and reduction. Climate security should not be viewed as any political movement in disguise. Scientific integrity should be respected. Additionally, Japan's international cooperation on policy study on climate security must be promoted to advance climate security policy in Japan.

Introduction

The Government of Japan released its new National Security Strategy (NSS) in December 2022, 9 years after the roll-out of its first-ever NSS in 2013. Today's international security landscape is completely different from that of 2013. It is duirng more recent years that climate change began to be discussed with a more aggressive expression such as 'climate crisis' in Japan as well.

Impact of climate change on our life is wide-ranging, including on health and sanitary conditions, economic activities, military operations and even national territory. It is not only a human security issue. The notion that climate change is a serious national security issue is growing among the Japanese today.[1]

Like other security challenges, climate security can be dealt with only by a whole-of-government approach and by international cooperation. The military forces of the world can work more closely to address the global challenge as a part of a whole-of-government approach. This issue provides the military forces of the world with a good opportunity to enhance cooperation. It could be regarded as a bright spot in the crisis.

The Ministry of Defense of Japan began to tackle this issue more seriously even during the middle of growing concern over the geopolitical threats posed by its belligerent neighbours (Ministry of Defense of Japan 2022a, 497–498). As Japan is surrounded by three nuclear weapon states, Russia, China and North Korea, the focus of its national security is naturally on traditional challenges posed by these countries. Nevertheless, awareness of the need to incorporate a climate security perspective into the whole picture of Japan's national security, was growing.

The basic tone of the new NSS is characterised by the great power competition, but cross-border issues including climate change, are not underrated. As the implementation of the new strategy has already started, it is important to keep watching how the climate security policy will be put into practice.

With this in mind, this article discusses how the Government of Japan, the Ministry of Defense in particular, should address climate security, particularly in close cooperation with international partners, focusing on the military aspect of climate security and based on the progress of climate change awareness in Japan.

Two notes are added to this point. First, while there may be room for military forces to reduce carbon dioxide emissions from their equipment and operations as they emit a considerable amount of greenhouse gas,[2] their main contribution to climate security is found in the areas of response and adaptation to climate change.[3] Therefore, this article discusses military-to-military cooperation mainly on the response and adaptation of the military forces to address a variety of impacts of climate change. Second, impact of the military issues on climate is debated today particularly due to the Russian invasion of Ukraine (e.g., Cohen 2022), and it is an important agenda to study, but this article focuses on the impact of climate change on the military as it is an issue of priority related to the natural disaster relief missions of the Japanese Self-defence Forces.

If Japan's defence establishment contributes to the improvement of the capabilities of its international partners to respond and adapt to the ramifications of climate change, it will contribute to the international efforts to build a security architecture particularly in the Indo-Pacific region by helping cement their security ties.

From 2013 NSS to 2022 NSS

Japan's 2013 NSS dealt with climate change as one of the global issues emerging as critical and urgent issues of human security, threatening the very survival and dignity of individuals. In accordance with the recognition that the global challenges could have repercussions on peace and stability of the international community, Japan committed to the promotion of necessary measures based on the principle of human security (Government of Japan 2013, 10). It also stated in terms of global economic risks, '[G]iven the aggravating environmental problems arising from climate change, there are risks of crunches in global supply and demand as well as temporary shortages of supply in food and water' (Government of Japan

2013, 10–11). It also expressed Japan's strong resolve to contribute to the achievement of emission reduction by the international community and to the resolution of climate change issues (Government of Japan 2013, 34).

International cooperation on disaster relief is also relevant to climate change. Regarding this point, the 2013 NSS expressed Japan's commitment to 'take the lead in international cooperation on disaster management and ensure that communities around the world have a high degree of resilience to disasters' (Government of Japan 2013, 33). This passage was added as the giant earthquake had hit Japan two years earlier. As Kameyama and Ono pointed out,

> 'Japan was more interested in this kind of disaster relief because disasters could happen even without any reference to climate change. Due to its experience with serious earthquakes, Japan was better able to understand the importance of resilience building, regardless of any connection to climate change' (Kameyama and Ono 2020, 276).

However, the scope of this passage was not limited to earthquakes. Though the 2013 NSS did not use the term 'climate security', the security document shows that the Government of Japan regarded climate change as a security issue.

Eight years after the release of the 2013 NSS, the Government of Japan started its review in 2021 in light of the sea change of the international security environment, including the heightened tensions between China and Taiwan. The 2022 NSS emphasises the necessity of military efforts for national security.

Assessment of the international security environment, both global and Indo-Pacific regional, is filled with geopolitical analysis. 'The world of a historical inflection point' (Government of Japan 2022b, 2) is not about the change of natural environment. Global issues, such as climate change, are touched upon in the section of the assessment only minimally. The 2022 NSS does not use the term 'climate security', either. However, it refers to the negative impacts of climate change in a number of other sections and declares, at the beginning of the independent section on 'Climate Change Measures' (Government of Japan 2022b, 16).

> 'Climate change is a security issue that affects the very existence of humankind. Extreme weather events due to climate change significantly impact Japan's national security in various ways including through more frequent and severe natural disasters, increased responses to disasters, more serious energy and food problems, a decrease in national land area, increased use of the Arctic sea routes'.

Odeyemi and Sekiyama pointed out that Japan explicitly started dealing with climate security as a policy issue in 2021 as 'a series of major events in the global arena likely caused a sudden heightened interest in climate risk in the 2020–2021 period in both the Japanese public and private sectors', referring to the influence of climate emergency declarations across the world around this period (Odeyemi and Sekiyama 2022, 14). According to them, 'it can be said that 2021 was the starting year for climate security discourse in Japan' (Odeyemi and Sekiyama 2022, 16). The review work of Japan's National Security Strategy started under this circumstance.

In terms of the negative impacts of climate change, it states, '[V]ulnerable countries in the Middle East, Africa, and the Pacific Islands are disproportionally suffering from extreme weather and reduction of land area due to climate change, the global spread of infectious diseases, and shortages of food and energy' (Government of Japan 2022b, 7). The section on 'Security Environment and Challenges in the Indo-Pacific Region' is just about geopolitical change and does not have any reference to climate change, except for a reference to 'natural disasters' in general as an example of 'threats and challenges of various types and intensities' (Government of Japan 2022b, 7). Nonetheless, it is a visible step forward as the 2013 NSS did not mention anything related to climate change in the section on the assessment of the security environment. It is also noteworthy that the 2022 NSS mentions 'Climate Change Measures' as one of the strategic approaches to achieve its security objectives.

Reference to 'food problems' shows greater emphasis on climate security in 2022 NSS than in 2013 NSS. 2013 NSS referred to the risk of food shortages due to the aggravating environmental problems arising from climate change, (Government of Japan 2013, 10–11) as quoted above. It also used the expression 'food security' (Government of Japan 2013, 10). 2022 NSS went further. It not only took up 'food problems' as one of the impacts of climate change on Japan's national security but also increased the description of food security (Government of Japan 2022b, 29–30). According to the study compiled by fishery scientists and economists, a large decrease in fish biomass is expected in the South China Sea due to the loss of suitable habitat in the face of worsening ocean temperatures. Many species reach the limits of their thermal tolerance range, and they seek more suitable habitats in cooler northern latitudes outside of the South China Sea (Sumaila et al. 2021, 18). Conflict over fishery resource is already becoming a real issue involving China and Japan. Japan's Fisheries Agency's patrol vessels ordered Chinese fishing boats to leave Japan's EEZ in the Sea of Japan 114 times in 2018, 1,115 in 2019 and 2,586 by the end of September 2020 (Furuya 2021). Food security aspect of climate change, particularly conflicts over fishery resource will have to be more highlighted in near future.

The section of 'Strategic Use of ODA and Other International Cooperation' also mentions climate change, stating, 'Also, under the concept of "human security," Japan will lead international efforts to solve global issues such as poverty reduction, health, climate change, environment, and humanitarian assistance' (Government of Japan 2022b, 17). Japan's efforts to address global challenges, including climate change, is based on the recognition that these issues 'have a direct or indirect impact on Japan's national security' (Government of Japan 2022b, 32–33).

The NSS is the overarching strategy document for the entire Government of Japan and is accompanied by the National Defense Strategy (NDS), which was released simultaneously with the 2022 NSS and has replaced the previous national defence policy called the National Defence Program Guidelines of 2018 (the 2018 NDPG). The NDPG was a cabinet decision to define the roles and missions of the Japanese Self-defence Forces and to establish the goal of the defence force build-up. The cabinet decision included the assessment of the security environment surrounding Japan as its basis. In contrast to the 2018 NDPG, which referred to Japan's proneness to natural disasters as part of the 'Security Environment Surrounding Japan' (Government of Japan 2018, 7), the NDS does not refer to it in the section of 'Changes in the strategic environment'. It focuses much more on geopolitical issues, but the section of 'Defence Challenges' mentions, 'Japan is prone to natural disasters that exact heavy damage' (Government of Japan 2022a, 9). Although not all natural disasters are caused by climate change, these descriptions are relevant to climate change. Frequent and large-scale natural disasters affect the activities of the Japanese Self-defence Forces. The impacts of climate change on the Japanese Self-defence Forces are not limited to the shift of operations. As in the section of 'Sustainability and Resiliency' of the NDS,

> 'the issue of climate change, including responses to future energy shifts, will inevitably further impact future MOD/SDF operations, including various plans, facilities, defence equipment and security environment surrounding Japan, which demand Japan to deal with a variety of issues related to these matters' (Government of Japan 2022a, 29).

Both the 2022 NSS and the NDS focus a lot on traditional security challenges, but obviously the Government of Japan recognises that climate change affects Japan's own national security. It proves that Japan's security mind-set is changing. The recent shift is thought to be owing to the following two factors in addition to the 'major events in the global arena' in the 2020–2021 period, which Odeyemi and Sekiyama pointed out (Odeyemi and Sekiyama 2022, 14).

First,

> 'based on recognition that climate change is a security challenge, the national defence organisations in various countries have widely grasped the climate change impact on security including the future energy shift issue, ..., while incorporating their ideas and policies to address this into their strategies and plans' (Ministry of Defense of Japan 2022b, 3–4).

The Ministry of Defense Response Strategy on Climate Change refers to the strategies and initiatives of the U.S., the UK, Australia and France (Ministry of Defense of Japan 2022b, 4). However, we have to pay attention to the fact that a vast amount of the climate security discussion has been lead by experts and practitioners worldwide for years, particularly since the UN Security Council discussed the matter in 2007 (Kameyama and Ono 2020, 271), and that it took more than a decade for the Japanese policymakers to begin their debate on this matter and establish the policy. It was a slow process. The Department of Defence of the U.S., i.e., Japan's ally, was addressing climate security issues in the era of the Obama administration and even of the Trump administration, but the Japanese Government did not pay attention to this policy trend of the U.S.

The sudden change of Japan's attitude in 2022 is attributed probably more to the second reason, that is, that the Japanese have noticed the large meteorological change in recent years. 'Typhoons and heavy rains have caused extensive damage including floods and landslides in a large number of areas (of Japan) and the advance of climate change is expected to cause even severer and more frequent disasters in the future' (Ministry of Defense of Japan 2022b, 6), which will have a direct impact on the central government, including the Self-defence Forces, and also on the local governments all over Japan.

In order to understand the points related to climate change in the 2022 NSS and the NDS, the policy proposal of 2022 by the Liberal Democratic Party (LDP), one of the two ruling parties, is helpful. It was released to the public eight months before the roll-out of the 2022 NDS in order to input ideas to the preparatory work of the strategy documents. The 16-page proposal has a section on climate change (LDP 2022, 16). Based on the recognition of the serious influence of climate change on Japan's national security, the LDP proposed the following three points. First, Japan should play a leading role in the international community for the resolution of climate-related global challenges and work with respective countries to assume their due responsibilities. Second, Japan should ensure the security of sea-lanes in the vicinity of Japan and the safety of Japan-related vessels sailing through the Arctic sea routes. And third, Japan should keep studying the posture to ensure sustainability of training and missions of the Self-defence Forces and enhance the resilience of the Forces' facilities and installations. The first point is incorporated into the NSS, the second and the third points are incorporated into the NDS.

The 2022 NSS lists 'climate change measures' in diplomatic efforts as part of the strategic approaches to attain the national security objectives, and entails international cooperation to address climate change, such as 'assistance to island nations and other developing countries where climate change poses imminent threats so that sustainable and resilient economies and societies can be built' (Government of Japan 2022b, 17). In contrast, the NDS does not specifically mention international security cooperation on climate change, but the climate change strategy of the Ministry of Defense devotes a number of pages to international cooperation (Government of Japan 2022b, 20–23). As the Japanese Self-defence Forces have a good track record of HA/DR and capacity-building cooperation with developing countries in the region, it is expected that the Japanese military will expand its climate security cooperation to entail an exchange of views and information on decarbonisation and an energy shift of military equipment and facilities. Intelligence cooperation on the geopolitical risks that climate change brings, must be enhanced, too.

Utility and scope of the term 'climate security'

As noted above, Japan's strategy documents do not use the term 'climate security'. According to Kameyama and Ono, the Ministry of Environment of Japan received the UK's proposal to include 'climate and security' in an agenda of the UN Security Council in 2007 with great interest and established a committee to discuss the idea. The committee's final report noted that the use of the term 'climate security' could help elevate the discussion of the climate change problem to 'high politics' and give this agenda a high priority in international politics, but the Ministry of Environment did not continue using the term 'climate security' after the publication of the report. They argued that the Ministry of Environment could have been concerned about causing confusion amongst Japanese people who think of traditional national security when the term 'security' is used (Kameyama and Ono 2020, 275). The real issue, however, is not 'confusion'. Japanese intellectuals and politicians who want to look away from military affairs tend to include as many non-traditional and non-military issues as possible into security discussion in order to dilute military aspect of security.[4] Noticing this, the Government of Japan seems to be generally reluctant to broaden the scope of the term 'security'.

While it is more and more relevant for today's world to discuss climate change in the context of national and international security, careful consideration is necessary in the use of the term 'climate security'.

As a common noun, 'security' means freedom from danger or threat (Oxford English Dictionary n.d.), but as dangers or threats are diverse and the way to reduce or eradicate them is also diverse, the scope of the term 'security' is elusive. The most important and difficult part of the security policy is answering the question of 'free from what'. It is about threat perception. If the change of the climate is regarded as a threat, reduction of greenhouse gas is regarded as the core issue of the security policy. In addition, 'security' is a convenient expression to express the seriousness, urgency and priority of any issue just in one striking word (Kameyama and Ocean Policy Research Institute 2022, 26). If so, those who use the expression 'climate security' must take the following two points fully into consideration:

First, as the term 'security' is a politically convenient word, people tend to use this term for whatever they regard critically important such as food, energy and economy. However, if the term is misused, it will lead to dilution of the impact of the meaning of security. It will not contribute to the solution of each of those diverse issues.

Secondly, as noted above, those who dismiss the military aspect of security tend to define 'security' as broadley as possible. However, the theory of realism says that security without military consideration is unreal in the anarchical international community (Kamiya 2018, 4). Defense of natural environment does not substitute for defence of national sovereignty. In fact, both efforts are necessary, but the latter would be easily dismissed by a wider definition of the term.

However, we should not deny the concept of 'climate security' as climate change has been affecting national and international security in its traditional sense. Ramifications of climate change are unprecedented. Conventional wisdom may not be useful any longer. Incidents caused by climate change affect the survival of sovereign states and their nationals. It is important to muster the efforts of all relevant organisations and international players. More emphasis must be added to these efforts.

It is a 'security' issue. The point is the scope of the term. This article focuses mainly on the response and adaptation of the military to the ramifications of climate change. Environmental protection is not be regarded as part of 'security'. It should be dealt with as part of environmental policy. In this article, reduction of carbon dioxide emissions is discussed only in terms of the emissions from military equipment, facilities and installations.

Evolution of climate security awareness in the Government of Japan in recent years

As noted above, the 2018 NDPG referred to Japan's proneness to natural disasters as part of 'Security Environment Surrounding Japan'. It originates in the 2004 version of the NDPG. Japan's proneness to be afffected by natural disasters is not new, but the reference to the fact in the NDPG indicates that the Government of Japan regards natural disasters, such as, typhoons, heavy rains and heavy snowfalls as security concerns. It is a natural expression of common sense of the ordinary Japanese as they suffer a lot from large-scale natural disasters every year. In fact, large-scale natural disasters are not limited to those of geological nature. In response to natural disasters, tens to even hundreds of thousands of troops are engaged in relief operations. HA/DR mission, particularly domestic large-scale natural disaster relief, is posited as one of the main missions that the Japanese Self-defence Forces should assume.[5]

As already noted, the present NDS does not refer to global issues such as climate change or Japan's proneness to natural disasters at all in its assessment of Japan's strategic environment. However, this does not mean that the Japanese military organisation treats climate change issues lightly. In fact, the recent efforts of the Ministry of Defense are noteworthy. There are a number of recent developments as follows:

First, in April 2021, the then Defence Minister Nobuo Kishi attended the Climate Security session in the virtual Leaders Summit on Climate, which US President Joe Biden convened, and presented the Ministry's efforts on climate change. He discussed the downward spiral risks of climate change threatening the peace and stability of the international community and emphasised the increase of disaster relief operations and close communication with

Southeast Asia and Pacific Island countries to address the impacts of climate change (Ministry of Defense of Japan 2021a).

Secondly, the Defence Ministry established the Climate Change Taskforce in May 2021 to analyse the impact of climate change on Japan's security and to take necessary measures (Ministry of Defense of Japan 2021c, 490). Compatibility of national defence with climate security is viewed as an important point in the discussion by the taskforce from its outset (Ministry of Defense of Japan 2021b). The taskforce has held six sessions so far,[6] and the fifth session on 29 August 2022 approved the Ministry's first-ever 'Ministry of Defense Response Strategy on Climate Change'.

Thirdly, the strategy document on climate change sets forth the following three objectives: one, 'the Ministry of Defense and the Self-defence Forces will adapt and respond appropriately to them (anticipated future impacts of climate change), and will fulfil the missions and roles in the future'; two, 'in order to achieve carbon neutrality by 2050, the Defense Ministry and the Self-defence Forces will strategically and steadily promote new initiatives with an eye to the future and with an awareness of timeline'; and finally, 'the Ministry of Defense and the Self-defence Forces will … aim to reduce total greenhouse gases emitted by the Ministry and the Self-defence Forces by 50% (excluding defense equipment) from the FY2013 level by FY2030' (Ministry of Defense of Japan 2022b, 1–2). The measures to be promoted in accordance with the Strategy include the following ten points: one, 'reinforcement of bases and other facilities and infrastructures to disasters'; two, 'improvement on defense capabilities and enhancement on resilience of defense equipment', including adaptation to energy shift; three, 'strengthening of sustainability and resilience in the logistics sectors'; four, 'reinforcement of disaster response capabilities'; five, 'reinforcement of strategic security cooperation'; six, 'improvement of living and working environments of Self-defence Forces personnel and reinforcement of sanitary functions'; seven, 'improvement of efficiency and reduction of greenhouse gas emissions at bases and other facilities'; eight, 'training, education and human resource development'; nine, 'reinforcement of (climate-related) technology base'; and finally, 'collaboration with local communities' hosting military bases (Ministry of Defense of Japan 2022b, 13–28).

Though not necessarily ambitious, this document deserves more recognition as the first-ever comprehensive account of the Defense Ministry's attempt to address climate change. Of particular importance is the following message:

> 'As adapting and responding to climate change, and maintaining and strengthening defence capabilities are compatible, the Ministry of Defense and the Self-defence Forces will simultaneously respond to both, as described in the following basic approach. Rather than considering the response to climate change negative for our country's defence, we see it as an opportunity to make our facilities and equipment more robust, resilient, and efficient for the future, and aim to simultaneously address climate change, and to maintain and strengthen our defense capabilities' (Ministry of Defense of Japan 2022b, 10–11).

Without a clear recognition of the compatibility with national defence, climate security would not be promoted by the Ministry of Defense.

Fourthly, Japan's Defence White Paper 2021 had one independent section on climate change for the first time in the history of the Defence White Paper, and spent three pages for the section on the description of the international security environment. Referring to the discussion in the Leaders Summit on Climate, the section concluded,

> 'During the discussions, defence officials noted that their ministries are increasingly called upon to respond to disasters thus elevating the need for enhanced disaster preparedness and response. In addition, they described the benefits of collaboration between defense ministers on shared climate crisis. As the defence authorities consider their measures and begin tackling them, climate change requires close monitoring with keen attention as a security issue' (Ministry of Defense of Japan 2021c, 201–203).

The 2022 edition of the Defence White Paper also allocates three pages to the section on the same topic. The section concludes,

> 'It is strictly required for the international society to unify their efforts against climate change, which is in progress worldwide and is even referred to as a threat multiplier. It is also needed to pay close attention to future efforts against climate change with great interest' (Ministry of Defense of Japan 2022a, 187–189).

The 2022 edition also refers to the Defense Ministry's commitment to respond to climate change while simultaneously maintaining and strengthening the defence capabilities, and emphasises the Ministry's promotion of the balance of renewable energy with national security (Ministry of Defense of Japan 2022a, 497–498). The newly published 2023 edition emphasises the attention of major countries' military organisations paid to climate security, saying that climate change is recognised as a high-priority issue by all countries, and the defence ministries and armed forces directly and indirectly affected by climate change have demonstrated their commitment to address the issue at all levels (Ministry of Defense of Japan 2023, 208).

Fifth and finally, the Defense Ministry plans to reduce greenhouse gas emissions, including by using renewable energy electricity. The Ministry of Defense/Self-defence Forces is the largest consumer of electricity in the Government of Japan, accounting for approximately 40% of total government consumption (Ministry of Defense of Japan 2022a, 498). The Ministry will 'work to ensure that 60% or more of the electricity procured by the Ministry of Defense and the Self-defence Forces will be generated by renewable energy sources by 2030' and 'aim to install photovoltaic power generation equipment in approximately 50% or more of Ministry of Defense and the Self-defence Forces buildings (including sites) that can be built by FY 2030' (Ministry of Defense of Japan 2022b, 24). It should be noted that the Ministry of Defense Response Strategy aims both energy efficiency and reduction of greenhouse gas emissions. Though Kameyama and Ono had argued in 2020, 'among Japanese policymakers, the climate change problem was perceived as an energy-saving problem and not as a security-related issue (Kameyama and Ono 2020, 275), it is not the case any longer.

Japan's defence diplomacy for climate security

Observers of Japan's national security policy tend to emphasise its dramatic transformation, reading the 2022 NSS and the NDS, but as stated in the 2022 NSS, the dramatic transformation is only about 'its execution' and 'the fundamental principles of the Japanese national security' are maintained (Government of Japan 2022b, 3). Japan's own military capability, a robust Japan-US Alliance and security cooperation with like-minded partners, will continue to form the main pillars of Japan's national security strategy. The more highlighted a whole-of-government approach is, as it should be, the brighter light should be shed on the enhancement of international security cooperation. Climate security cooperation deserves greater attention simply because it is a pressing issue that can be addressed only in global concert.

When the Government of Japan explains its national security and defence policy, it usually bifurcates international security cooperation, particularly military cooperation, into Japan-US Alliance cooperation and security cooperation with other like-minded countries as two separate pillars of the policy (Government of Japan 2022a, 10), but this distinction is becoming blurred as is reflected in the arrangement of 'ally' and 'like-minded countries' in a row in a number of places in the 2022 NSS and the NDS. Accordingly, this article deals with the international security cooperation without such distinction.

Japan's international security cooperation has a number of tools, from personnel exchange, unit-level cooperation, to military equipment and technology cooperation (Ministry of Defense of Japan 2022a, 327). The Japanese Self-defence Forces should use these tools effectively and proactively for security exchange and cooperation with its international partners bilaterally and in multilateral frameworks, such as, the US-centred alliance network in the region, the Quad, the ARF and the ADMM+. Information sharing is always helpful as the basis of capacity building and interoperability. Unit-level operational cooperation such as exercise is instrumental for the enhancement of capacity. Equipment and technology cooperation can directly help enhance partners' technological capacity.

The 2022 NSS lays a political foundation for climate security cooperative efforts. As it is quite new for Japan and urgent, there is much room for expansion.

Cooperation in *response*

Major candidate areas for cooperation between the Japanese military and its international partners in order to *respond* to the situations generated by climate change, will be HA/DR and Arctic security.

HA/DR cooperation

Disaster relief is one of the main missions of Japan's Self-defence Forces. The expectations of the Japanese nationals on the disaster relief operations of the Self-defence Forces are always high. Utilising the expertise on disaster relief in Japan, the Self-defence Forces have been engaging in HA/DR missions overseas and in capacity building on HA/DR. When *Super Typhoon Haiyan* hit the Philippines in November 2013, Japan's Self-defence Forces formed their first-ever joint task force for relief operations. The operation was the largest they had ever conducted with around 1,100 troops. This region is more likely to be hit by a similar or even larger disaster, and HA/DR cooperation will be more important to maintain regional stability. Large-scale natural disasters will easily cause deterioration of public order.[7] Disaster-hit countries may become unguarded and vulnerable against foreign threats. Disinformation will be easily spread in the disaster-hit areas and foreigners easily become victimised by disinformation, which will cause international tension. In order to facilitate closer cooperation in HA/DR, it is important to make plans, create procedures and to enhance joint exercises to improve their skills.

The Pacific Island nations are important partners for Japan in HA/DR cooperation. In September 2021, the Defense Ministry virtually held the Japan Pacific Islands Defence Dialogue (JPIDD). This was the first-ever multilateral defence ministerial dialogue hosted by the Ministry. The Pacific Island region has a strong historical relationship with Japan, and today, the geostrategic importance of the region is increasing. The 2018 NDPG referred for the first time in the history of Japan's defence policy to Japan's intention to promote cooperation with those countries (Government of Japan 2018, 17). The JPIDD Joint Statement stressed the significance of enhancing cooperation among defence authorities on disaster risk reduction, especially for building resilience, and emphasised the benefits of effective coordination among defence and relevant authorities to enhance the provision of HA/DR in times of crisis triggered by natural disasters in the region (Ministry of Defense of Japan 2021d). HA/DR will continue to be a main issue for cooperation with the regional countries.[8] The Defense Ministry's Response Strategy on Climate Change lists, as one of the measures for strategic security cooperation, the effort to implement and enhance seminars and capacity building support by making effective use of lessons learned and know-how accumulated so far by the Ministry of Defense and the Self-defence Forces through domestic and international disaster relief operations, HA/DR, and other activities while maintaining close communication with ASEAN countries and Pacific Island countries (Ministry of Defense of Japan 2022b, 22). Addressing sea level rise due to climate change is another urgent issue for security cooperation. In low-lying countries whose lands are only a few metres above sea level, most of the populations are at risk from rising sea levels. In the Maldives and the Marshall Islands, sea level rise is predicted to affect more than 40% of the population. In Bangladesh, up to 20 million people might move to other parts of the country or to other countries by 2050 due to rising sea levels and floods, and these environmental immigrants could become a source of conflict within Bangladesh or in India and other neighbouring countries (Sekiyama 2002, 53).[9] The 2022 NSS also mentions the Pacific Island countries as those 'disproportionally suffering large damages from extreme weather and reduction of land area due to climate change' (Government of Japan 2022b, 7).

Arctic security

Global warming melts Arctic sea ice, which potentially opens new trade routes in international waters in the Arctic Ocean (Lynch, Norchi, and Li 2022). Former Defense Minister Kishi said in the Climate Security session of the *Leaders Summit on Climate* in 2021 that if we take an Arctic sea route from Japan to Europe, the distance can be reduced to 60% of the existing route *via* the Suez Canal (Ministry of Defense of Japan 2021a, 2022b, 21). Not only the eight Arctic States but also other players such as Japan, China and India are more interested in the Arctic.

China called itself a 'Near-Arctic State' in the white paper on its official Arctic policy in 2018 (Wong 2018), though such a category as 'Near-Arctic State' does not exist. According to the Chinese white paper, China wants to tie the Arctic to its Belt and Road Initiative, encouraging joint efforts to construct a 'Polar Silk Road' that would link China and Europe through the Arctic Ocean (Wong 2018).

India is also increasingly interested in the Arctic. It says that India's engagement with the Arctic started in 1920 (Government of India 2022, 5), more than one hundred years ago. India is concerned about the impact of the change in the Arctic, especially melting Arctic ice on the country's national development. India's Arctic Policy states,

> 'Studying the impact of climate change in the Arctic can improve response mechanism in other parts of the globe. The atmosphere is warming faster in the Arctic, whereas the ocean is warming faster in the Indian Ocean. It is therefore imperative to understand causative mechanisms of such warming and predict outcomes' (Government of India 2022, 13).

Although the Defence Ministry of India does not seem to be officially involved in India's Arctic engagement (Government of India 2022, 8), the country's national security considerations seems to be relevant to the Arctic Policy as promotion of security and stability in the Arctic region in accordance with relevant international law, is among the objectives of India in governance and international cooperation involving the Arctic (Government of India 2022, 21).

The Arctic will be an important route for the flow of military forces, too. Sea-lane defence in the Arctic Ocean will be a matter of interests for a number of countries. The major powers' increasing attention to the region is likely to bring their competition, including their respective militaries, to the region.

Japan's Defense Ministry is

> 'concerned that it (the melting of sea ice) will have an impact on Japan's security such as instability due to disputes among major powers and related countries over the resources of the Arctic Ocean, China's advancement into the Arctic Ocean *via* the Sea of Japan and the passage becoming an important shipping route' (Ministry of Defense of Japan 2022b, 8).

The strategic document is not necessarily specific about possible roles of the Japanese Self-defence Forces on Arctic security, just saying,

> '[W]e will also cooperate with relevant ministries and agencies, and countries, etc., to take serious interest in the impact of climate change on Japan's security environment as a security issue, conduct wide-ranging information gathering and assessment, and make efforts to share a common understanding and awareness of the threats, challenges, and countermeasures of climate change in collaboration with relevant ministries and agencies, relevant countries and other authorities on an issue-by-issue basis' (Ministry of Defense of Japan 2022b, 21).

HA/DR, search and rescue, and sea-lane defence[10] in the Arctic region will be in the scope of international security cooperation.

However, seeing the declaration of the seven Arctic countries to suspend cooperation with Russia at the Arctic Council due to the Russian invasion of Ukraine, some experts express another concern, 'although a conflict in the Arctic may be unlikely, it's hard to gauge whether it would happen, based on Russian President Vladimir Putin's unpredictable behaviour and his invasion of Ukraine' (Sacks and Van Abel 2022). If so, joint efforts to deter armed conflicts will be necessary in the Arctic, but the possibility of the Arctic for more frequent commercial shipping should not be denied as global warming in the Arctic is progressing twice as fast as in other regions of the world and the Arctic is predicted to start seeing iceless summers within the next ten years and sea ice in the region will completely disappear during summer by the middle of this century (Sekiyama 2002, 58).

Cooperation in *adaptation*

Major candidate areas for cooperation in order to *adapt* to the situations generated by climate change will be military medicine (tropical medicine in particular), training assistance and exchange of expertise on military operations in extreme heat. Global warming affects performance of military service members in combat and in training. It also affects military facilities. Expertise in these areas should be shared among security partners to enhance their capabilities and to avoid unnecessary risks.

The climate change response strategy of Japan's Defense Ministry commits to the following measures: to 'promote examining clothing and equipment taking into account the increased health risks for Self-defence Force personnel who are expected to operate in heat waves and under extremely high temperatures, as well as the need for air conditioning equipment for defense equipment'; to 'work to secure medicines and other resources for diseases'; to 'work to address vector control (mosquitoes, rodents, ticks, etc.) at bases and other facilities' (Ministry of Defense of Japan 2022b, 23); and 'training, education and human resources development' (Ministry of Defense of Japan 2022b, 26). These measures will help conduct and enhance cooperation with international partners on their adaptation to climate change.

Military medical cooperation has become an area for international cooperation since the spread of Covid-19, and the channels of communication for the cooperation and accumulation of expertise will help enhance military medical and health cooperation related to climate change.

Cooperation in *reduction*

The military forces will be able to and should contribute to the *reduction* of greenhouse gas emissions. Thus, exchange of expertise on how to reduce greenhouse gas emissions from equipment will be a possible area for military cooperation.

Whether reduction of greenhouse gas emissions from the military is considered as a matter of security policy, it will not be conceivable for the military to be exempted from the efforts for the reduction anyway. There is a view that as the Self-defence Forces accounts for only about 0.5% of Japan's total oil consumption, merits of the use of biofuels will be limited (Cabinet Secretariat of Japan 2022, 39), but if most of the vehicles for civilian use become driven with fuels other than fossil ones, the military use of fossil fuel vehicles will be extremely costly. Rather, the Japanese military establishment should think of how to achieve compatibility between green effort and military capability build-up. This is not a challenge solely for Japan. Other countries' efforts are already in progress. For example, one of the guiding principles of the strategic approach of the UK's Ministry of Defence on climate change is operational capability. The Defence Ministry explains, 'Defense must preserve a capability despite the implications of climate change', and 'Defense will take opportunities to enhance capabilities and resilience with new options while never compromising capability solely for a sustainable solution' (Ministry of Defence of the UK 2021, 11). For another example, Singapore's Defence Science and Technology Agency (DSTA) signed a Memorandum of Understanding with ExxonMobil to explore the use of sustainable fuel. According to DSTA, the collaboration will also allow it to better understand ways to reduce emissions without compromising naval vessels' operational effectiveness (Defence Science and Technology Agency 2023, 1).

Military planners should be aware that the use of renewable energy has a potential to improve military operational effectiveness. For example, if solar energy is used for military equipment, it may reduce the burden of logistics for fuel transportation. As defence of natural environment is not a mission they regard as theirs, it is not easy to give them enough motivation to work on it. The statement in the Defense Ministry's Climate Strategy that 'adapting and responding to climate change, and maintaining and strengthening defence capabilities are compatible' (Ministry of Defense of Japan 2022b, 10) is good, but it is necessary to gain the momentum to put the recognition into practice. Defence practitioners will have little interest in reducing greenhouse gas emissions just for the purpose of reduction, and thus it is critically important to convince them that it will contribute to their defence capability and efficiency. Just labelling the reduction effort as a 'security' issue will not be enough.

Relying on renewable energy must require caution. Japan's 2022 NSS cites climate change as an issue that require cross-border cooperation among nations (Government of Japan 2022b, 1). However, measures to address climate change are affected by geopolitical rivalry. Materials used in key renewable energy technologies are potentially weaponised in the geopolitical competition. For example, China currently controls more than half the global processing capacity for many of these materials, including rare earths for electric vehicle motors; polysilicon for solar panels, and cobalt, lithium manganese and graphite for electric vehicle batteries (National Intelligence Council of the USA 2021, 6). International cooperation for the resilience of the supply chain of these key materials is also necessary.

Cooperation in arms control

One more issue may be added to the list of international security cooperation related to climate change. It is about an agreement of 1977 called *Convention on the Prohibition of Military or Any Other Hostile Use of Environmental Modification Technique*. As of 21 August 2023, the Convention has 78 participating countries, including Japan (United Nations 2023). In light of the history that rain-making technology was used in the Vietnam War, military use of environmental modification techniques should not be considered just as a fiction. The U.S. forces conducted rain-making operations in Southeast Asia from 1967 to 1972 in an attempt to slow the movement of North Vietnamese troops and supplies through the Ho Chi Minh Trail network. The cloud-seeding program is said to have increased local rainfalls up to 30% (Hersh 1974, 1). While it is debateable if the military program modified the environment or not, and it is not clear in the language of the Convention which technology is specifically banned, there was a news report just before the Beijing Winter Olympic that China has long attempted to modify the weather through cloud-seeding, a process of trying to stimulate rainfall by firing silver iodide-filled shells and rockets into clouds, that a large number of people work for the Beijing Weather Modification Office and other provincial counterparts and that China was likely to use the technology to make the Winter Olympic successful (Zeitchik 2022). This Convention bans military and hostile use of environmental modification techniques alone, and it may not be relevant to efforts to address greenhouse gas emissions, but should not be dealt with separately. The international community should pay much more attention to this international agreement and the contracting parties should discuss the interpretation and implementation of the convention more intensely.

For fruitful cooperation towards the future

Climate science is often politicised. While a sceptic of the theory of climate change discusses the danger of politicised science (Appendix I 'Why Politicised Science is Dangerous', in

Crichton 2004), those who argue against climate change theory are also politicising the science. Climate change and other environmental issues are often viewed as favourite agenda items for 'liberals' and they are not favoured by 'conservatives', but action against climate change is not a leftist movement or any other political movement in disguise. Politicisation of the issue is somewhat inevitable, but it will not necessarily help generate a great momentum based on public consensus to promote measures to address climate change. Rather, cultivation of scientific integrity should be the first priority.

Although all security issues are not responsibility of the military forces, the public and even experts tend to look to military experts first when addressing any 'security' issues including climate security. Therefore, it is important and instrumental, first and foremost, to gain the understanding of the military experts on all the major issues related to climate change.

Climate security is one of the non-traditional and trans-national security issues for which international cooperation based on a whole-of-government, a whole-of-alliance and a whole-of-partnership approach, including military contribution, is indispensable. Menu of Japan's international cooperation for climate security is multiple. Exchange of views and information and sharing a common picture of the entire issue is the basis of practical cooperation. Capacity building and technology cooperation is instrumental for strengthening individual and joint capabilities to address climate change. Operational cooperation such as HA/DR and its exercise is necessary to respond to actual crises and to prepare for possible crises. Arms control cooperation on military and hostile application of geo-engineering technologies will help generate a stable and predictable international order. Also, it is expected that if international cooperation for climate security is enhanced, it will spill over to other types of security cooperation and contribute to regional security networking, including Southeast Asia and the Pacific Island region. International cooperation to address climate change is too important to be overshadowed by geopolitical rivalry.

Finally, the Japanese policymakers must pay attention to the fact that they were indifferent to the policy trend on climate security in the international community until recently (Kameyama and Ono 2020, 271). On this point, Odeyemi and Sekiyama pointed out that the scarcity of systematic research on climate security has affected policy progress in Japan (Odeyemi and Sekiyama 2022, 10). If so, Japan's international cooperation on policy study on climate security must be promoted to advance climate security policy in Japan.

Disclosure statement

No potential conflict of interest was reported by the author(s).

Notes

1. Kazumine Akimoto argues that if the Arctic sea routes had been open to the Russian Baltic Fleet at the time of the Russo-Japanese War in the early twentieth century, the outcome of the war might have been different (Akimoto 2022a, 9).
2. There is an estimate that the world's total military carbon footprint is approximately 5.5% of the global total, and that if the global militaries were a country, they would have the world's fourth largest footprint, one larger than whole of Russia (Parkinson and Cottrell 2022, 9-10).
3. Kazumine Akimoto refers to five points as the main foci of climate security efforts related to sea areas, including rescue activities and dealing with the impacts on national defense functions (Akimoto 2022b, 39-40). Japan's White Paper on Defence emphasizes shifts of missions of the military establishments and their adaptation in terms of logistics, equipment and human resources much more than their greenhouse gas emission (Ministry of Defense of Japan 2022a, 187).
4. According to Matake Kamiya, the concept of 'comprehensive security' developed in post-WWII Japan, but reflecting the Japanese nationals' cautious views on the military, the 'comprehensive' security concept in Japan tended to unfairly dismiss military aspects of security or even consider it as a crime (Kamiya 2018, 21).
5. See Article 83, Sefl-Defense Forces Law.
6. The sixth session was held on July 10, 2023.
7. Sekiyama refers to empirical study which states that floods may prolong civil wars (Sekiyama 2002, 46).
8. Ministry of Defense Response Strategy on Climate Change states, 'the Self-Defense Forces is expected to play an increasing role in capacity building support, HA/DR, etc. to these countries (the Pacific Island countries)' (Ministry of Defense of Japan 2022b, 8).
9. In Japan, Okinotorishima Island is an atoll like the Maldives and the Marshall Islands. It is predicted to be submerged in the middle of the 21st century at earliest due to sea level rise, and Japan will lose the EEZ surrounding the island (Kayane 2009). Also in Japan, 46% of the population live in the local municipalities facing the sea, and 47% of industrial production come from those areas (Ministry of Environment of Japan 2001). Therefore, sea level rise will seriously affect Japan's own territorial integrity and economic prosperity as well, and this issue is becoming a national security issue for Japan, too.
10. If the international community becomes less dependent on fossil fuels, then energy transport routes may be changed. If so, the Indian Ocean and the Arabian Sea might become less important. Instead, the South Pacific routes connecting Australia with the rest of the world, for example, might be much more important if Australia's cheap hydrogen production is taken into consideration.

References

Akimoto, K. 2022a. "The Security Environment Created by Climate Change." In *Climate Security: Global Warming and a Free and Open Indo-Pacific*, edited by Ocean Policy Research Institute, 3–9. Tokyo, Japan: Tokai Education Research Institute.

Akimoto, K. 2022b. "Efforts to Address Climate Security in the Indo-Pacific Region: Making Use of Essential Defense Functions." In *Climate Security: Global Warming and a Free and Open Indo-Pacific*, edited by Ocean Policy Research Institute, 31–40. Tokyo, Japan: Tokai Education Research Institute.

Cabinet Secretariat of Japan. 2022. "Aratana Kokkaanzenhoshousenryakutouno Sakuteinikansuru Yuushikishatono Ikenkoukan (Gironno Youshi) [Exchange of Views with Experts on the Establishment of a New National Security Strategy and Related Policies (Summary of Discussions)]." https://www.cas.go.jp/jp/siryou/pdf/yousi.pdf

Crichton, M. 2004. *State of Fear*. New York: Avon Books.

Cohen, S. 2022. "The Impact of Russia's Invasion of Ukraine on Climate Change Policy." Columbia Climate School, March 7, 2022. https://news.climate.columbia.edu/2022/03/07/the-impact-of-russias-invasion-of-ukraine-on-climate-change-policy/

Defence Science and Technology Agency. 2023. "DSTA Signs MOU with ExxonMobil to Explore Use of Sustainable Fuel."

Furuya, K. 2021. "A New Threat in the Sea of Japan – Chinese Fishing Boats and Resource Depletion." *International Information Network Analysis*, February 1, 2021. https://www.spf.org/iina/en/articles/furuya_02.html

Government of India. 2022. "India's Arctic Policy – Building a Partnership for Sustainable Development." https://www.moes.gov.in/sites/default/files/2022-03/compressed-SINGLE-PAGE-ENGLISH.pdf

Government of Japan. 2013. "(Provisional Translation) National Security Strategy." https://www.cas.go.jp/jp/siryou/131217anzenhoshou/nss-e.pdf

Government of Japan. 2018. "(Provisional Translation) National Defense Program Guidelines for FY 2019 and beyond." https://www.mod.go.jp/j/approach/agenda/guideline/2019/pdf/20181218_e.pdf

Government of Japan. 2022a. "(Provisional Translation) National Defense Strategy." https://www.mod.go.jp/j/policy/agenda/guideline/strategy/pdf/strategy_en.pdf

Government of Japan. 2022b. "(Provisional Translation) National Security Strategy of Japan." https://www.cas.go.jp/jp/siryou/221216anzenhoshou/nss-e.pdf

Hersh, S. 1974. "U.S. Admits Rain-Making from '67 to '72 in Indochina." *New York Times*, May 19, 1974.

Kameyama, Y. 2022. "The Concept of Climate Security and Its Importance." In *Climate Security: Global Warming and a Free and Open Indo-Pacific*, edited by Ocean Policy Research Institute, 23–30. Tokyo, Japan: Tokai Education Research Institute.

Kameyama, Y., and K. Ono. 2020. "The Development of Climate Security Discourse in Japan." *Sustainability Science* 16 (1): 271–281. https://doi.org/10.1007/s11625-020-00863-1

Kamiya, M. 2018. "Anzen-Hosho-No Gainen [Concepts of Security]." In *Shintei-Dai-5-Han Anzen-Hosho-Gaku Nyumon [Introduction to Security Studies]*, edited by Boei-daigakko anzen-hosho-gaku ken-kyukai, 5th ed., 3–27. Tokyo, Japan: Aki Shobo.

Kayane, H. 2009. "Kaimen-josho-ni-taisuru Okinotorishima-no iji [Maintenance of Okinotorishima Island against sea level rise]." University of Tokyo, July 20, 2009, https://www.oa.u-tokyo.ac.jp/researcher-story/012.html

Liberal Democratic Party (LDP). 2022. "Aratana Kokkaanzenhoshousenryakutouno Sakuteinimuketa Teigen [Proposal for the Establishment of a New National Security Strategy and Related Policies]." Liberal Democratic Party. https://storage.jimin.jp/pdf/news/policy/203401_1.pdf

Lynch, A., C. Norchi, and X. Li. 2022. "The Interaction of Ice and Law in Arctic Marine Accessibility." *Proceedings of the National Academy of Sciences of the United States of America* 119 (26): 1–3. https://doi.org/10.1073/pnas.2202720119

Ministry of Defense of Japan. 2021a. "(Provisional Translation) Minister of Defense Mr. Nobuo Kishi Attended the Climate Security Session in the Leaders Summit on Climate." https://www.mod.go.jp/en/images/c1e9427e9c3242ddc417bb4acddecd9202775be8.pdf

Ministry of Defense of Japan. 2021b. "Dai 1-kai Boueisho Kiko-hendo Taskforce Giji-gaiyo [Summary of the First Meeting of Climate Change Taskforce of the Ministry of Defense." https://www.mod.go.jp/j/approach/agenda/meeting/kikouhendou/pdf/gijigaiyo_01.pdf

Ministry of Defense of Japan. 2021c. "Defense of Japan 2021."

Ministry of Defense of Japan. 2021d. "Japan Pacific Islands Defense Dialogue Joint Statement." https://www.mod.go.jp/en/images/e4b4d8b1ad510d146cab838b7f0a65ebbef9c99f.pdf

Ministry of Defense of Japan. 2022a. "Defense of Japan 2022."

Ministry of Defense of Japan. 2022b. "Ministry of Defense Response Strategy on Climate Change."

Ministry of Defense of Japan. 2023. "Reiwa 5-nenban Nippon-no Bouei [Defense of Japan 2023]."

Ministry of Defence of the UK. 2021. "Ministry of Defence Climate Change and Sustainability Strategic Approach." https://assets.publishing.service.gov.uk/media/605ddbbe8fa8f5047d3a851e/20210326_Climate_Change_Sust_Strategy_v1.pdf

Ministry of Environment of Japan. 2001. "Chikyu-ondanka-no Nihon-eno eikyo 2001 Gaiyo [Impact of Global Warming on Japan, 2001, Outline]." April 26, 2001. https://www.env.go.jp/press/files/jp/2223.html

National Intelligence Council of the USA. 2021. "National Intelligence Estimate: Climate Change and International Responses Increasing Challenges to US National Security through 2040."

Odeyemi, C., and T. Sekiyama. 2022. "A Review of Climate Security Discussions in Japan." *International Journal of Environmental Research and Public Health* 19 (14): 8253. https://doi.org/10.3390/ijerph19148253

Oxford English Dictionary. n.d. Accessed August 22, 2023. https://www.oed.com/search/dictionary/?scope=Entries&q=security

Parkinson, S., and L. Cottrell. 2022. "Estimating the Military's Global Greenhouse Gas Emissions." Scientists for Global Responsibility. https://ceobs.org/wp-content/uploads/2022/11/SGRCEOBS-Estimating_Global_MIlitary_GHG_Emissions_Nov22_rev.pdf

Sacks, B., and K. Van Abel. 2022. "How the Russian Invasion of Ukraine May Impact the Arctic." RAND Blog. https://www.rand.org/blog/2022/08/how-the-russian-invasion-of-ukraine-may-impact-the.html

Sekiyama, T. 2002. "Risks and Realities of Violent Conflict Caused by Climate Change." In *Climate Security: Global Warming and a Free and Open Indo-Pacific*, edited by Ocean Policy Research Institute, 43–59. Tokyo, Japan: Tokai Education Research Institute.

Sumaila, U. R., W. W. L. Cheung, L. S. L. The, A. H. Y. Bang, T. Cashion, Z. Zeng, J. J. Alava, S. I. Clue, and Y. S. de Mitcheson. 2021. "Sink or Swim: The Future of Fisheries in the East and South China Seas." Hong Kong: ADM Capital Foundation. https://www.admcf.org/wp-content/uploads/2021/11/Sink-or-Swim-Full-Report_171121.pdf

United Nations. 2023. "United Nations Treaties Collection." https://treaties.un.org/Pages/ViewDetails.aspx?src=IND&mtdsg_no=XXVI-1&chapter=26&clang=_en

Wong, A. 2018. "China: We Are a 'Near-Arctic State' and we want a 'Polar Silk Road'." *CNBC*, February 14, 2018. https://www.cnbc.com/2018/02/14/china-we-are-a-near-arctic-state-and-we-want-a-polar-silk-road.html

Zeitchik, S. 2022. "The Winter Olympics Could Highlight China's Innovative – and Troubling – Efforts to Control the Weather." *Washington Post*, January 24, 2022. https://www.washingtonpost.com/sports/olympics/2022/01/24/olympics-china-weather-control/

Index

Note: Figures are indicated by *italics*. Tables are indicated by **bold**.